NATIONAL AUDUBON SOCIETY® POCKET GUIDE

An Andrew Stewart Publishing Edition

John Farrand, Jr.
Natural Science Editor
Chanticleer Press, Inc.

Familiar Mammals

Alfred A. Knopf, New Yor[k]

This is a Borzoi Book
Published by Alfred A. Knopf

 Published in the United States
by Alfred A. Knopf, a division of Penguin Random
House LLC, New York, and distributed in Canada
by Random House of Canada, a division of Penguin
Random House Ltd., Toronto.

Color reproductions by Nievergelt Repro AG, Zurich,
Switzerland.
Typeset by Dix Type, Inc., Syracuse, New York.
Printed and bound by Toppan Leefung Printing
Limited, China.

Published March 1988
Seventeenth printing, October 2015

Library of Congress Catalog Card Number: 87-46022
ISBN: 978-0-394-75796-4

Contents

Introduction
How to Use This Guide 6
Identifying Mammals 8
Finding Mammals 10
Track Guide 12

The Mammals
Rabbits and Hares 18
Squirrels and Marmots 26
Pika 56
Rodents 58
Opossum 88
Armadillo 90
Moles 92
Pocket Gopher 96
Shrews 98
Bats 104
Carnivores 110
Hoofed Mammals 158

Appendices
Guide to Orders and Families 178
Index 186
Credits 190

How to Use This Guide

Mammal-watching, like bird-watching, is a rewarding pastime that will add much to your enjoyment of any trip outdoors. Mammals are common everywhere, and learning to find and identify them is fun and challenging, whether you are in a national park or in your own backyard.

Coverage

This guide gives full coverage to 80 of the most common and distinctive mammals in North America, from the Arctic to the Mexican border, and from the Atlantic coast to the Pacific. An additional 30 important mammals are included under the heading "Similar Species," raising to 110 the total number of mammals described in this guide.

Organization

This easy-to-use pocket guide is divided into three parts: introductory essays, illustrated accounts of the mammals, and appendices.

Introduction

Two introductory essays will help you to use and enjoy the guide. "Identifying Mammals" tells you what to look for when you see an unfamiliar mammal, and "Finding Mammals" gives practical clues about where to go and what to do in order to discover these often elusive animals.

The Mammals

This section includes 80 color plates, arranged visually by

shape, color, and overall appearance. Every major group of mammals except the whales and dolphins is represented by at least one species. Opposite each illustration, along with the mammal's English and scientific names, is a description of the species' important field marks, notes on other species with which it might be confused, and descriptions of its habitat and range. For quick reference, a map showing each species' distribution supplements the range statement. A diagram of the species' tracks is included when this is helpful in finding or identifying the mammal. An introductory paragraph provides additional information on each mammal's habits and life history.

Appendices

Following the species accounts, a special section "Guide to Orders and Families" describes the eight major groups of mammals covered in this book. Remembering the distinguishing features of these broad categories will often help you to identify a species more quickly and easily.

Knowing how to find and identify mammals, and learning about their habits, will enrich every excursion you make and increase your understanding of the natural world.

Identifying Mammals

Most of us know a deer, fox, or mouse when we see one. But each of these is a broad category of mammals that may contain several species. To identify a mammal more precisely—to name its species—you should check a number of features. With some mammals these features may be subtle, but for most just a brief glimpse is all you need. Whenever you see an unfamiliar mammal, you should note its size, shape, color and pattern, habitat, and range.

Size

Mammals show a great range in size, and it is usually easy to make a rough estimate of how big a mammal is. Comparing the size of an unfamiliar mammal to that of a mouse, cat, or large dog can be helpful.

Shape

Deer, seals, and bats have unmistakable silhouettes, but other groups, such as shrews and mice, are about the same size but differ in the shape of the snout, the length of the tail, and usually the size of their ears. Noting a mammal's shape will help you to narrow it down to a broad category—like rabbits or moles. Then you can pin it down to a particular species. Of course, in many cases the shape of a mammal will be all you need to identify it to species; nothing else looks like a Bison, Nine-banded Armadillo, or Porcupine.

Color and Pattern Skunks, chipmunks, the Raccoon, and a few other mammals can quickly be recognized by their colors and patterns. But most are clad in soft browns or tans. Any markings are easy to miss if you don't already know what to look for. So when you come across an unfamiliar mammal, make a point of noting and remembering as many details of color and pattern as you can. Reading the descriptions in this guide ahead of time will help you to spot important colors and patterns in the field.

Habitat Where you find a mammal can often help you decide what it is. A Mink, for example, is almost always found near water, and seldom or never climbs trees, but the similar Fisher is an expert tree-climber, and is often found far from ponds or streams. Be sure to check the habitat descriptions so you will know what to expect when you go out looking for mammals.

Range One of the easiest ways to eliminate possibilities when you are trying to identify a mammal is to check its range. If you are in California, for example, and learn that the mammal you suspect you have seen is only found in the Southeast, you can be certain it was something else.

Finding Mammals

Some mammals are easy to find, but most are nocturnal and secretive. Locating them takes patience and sharp eyes. When you find a mammal, few pleasures are as great as watching it in its natural habitat. You need not be in a remote wilderness to see mammals. State parks and even suburban areas are excellent places to look, because the animals there have grown accustomed to people and will probably not be as shy or difficult to approach. You may not bring home a long list the way bird watchers do, but what you have seen will be every bit as exciting.

Tracks

Before you go out looking for mammals, check the habitat descriptions in this guide so you will know which species to expect. Study the illustrations of tracks, because your first clue that a particular animal is around will often be a set of tracks on the muddy margin of a pond or in dry sand beside a road.

Time of Day

Although many mammals can be seen in broad daylight, the best times to go looking are at dusk or dawn, when they are just starting their night's activity or heading for home. That is when you are most likely to see a fox bounding across a field or a deer moving slowly along a forest path. After dark is also a good time, especially if you have a flashlight. Keep the beam of the flashlight parallel to your line of sight, so you can pick up the reflected eyeshine of nocturnal animals.

Observations

Mammals have keen senses. Their ears, noses, and eyes are always on the alert. When you are in the field, avoid doing anything that will attract attention. In a forest make sure your shoes don't rustle the dry leaves and your clothing doesn't scrape on branches. If you can hear yourself, the animals can hear you, too. Try to walk into the wind so your scent won't reach an animal before you see it. Mammals also notice motion, so move as little as possible.

One of the best ways to see mammals is to find a good spot and then sit down and wait. In a forest, sit near a trail or stream. In open country, use the shelter of a bush or rock and be sure you aren't silhouetted against the sun. Before long the mammals will ignore your silent figure and resume their normal activities. Once you are sitting quietly and have become part of the landscape, you will be surprised at how many mammals you will see and fascinated by what you see them doing. Nearly everything we know about mammals was learned by someone sitting still and watching them.

Track Guide

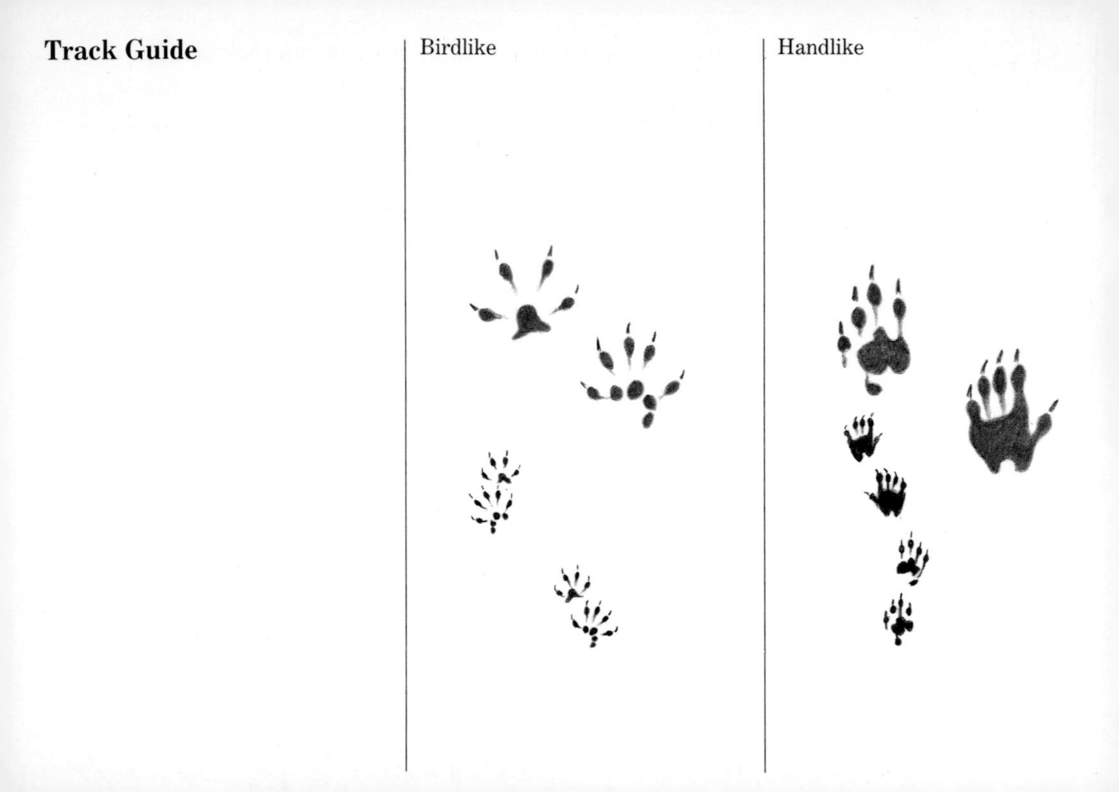

Hand- and Footlike
4-toed foreprints, 5-toed
hindprints

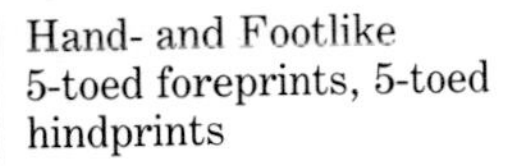

Hand- and Footlike
5-toed foreprints, 5-toed
hindprints

Footlike

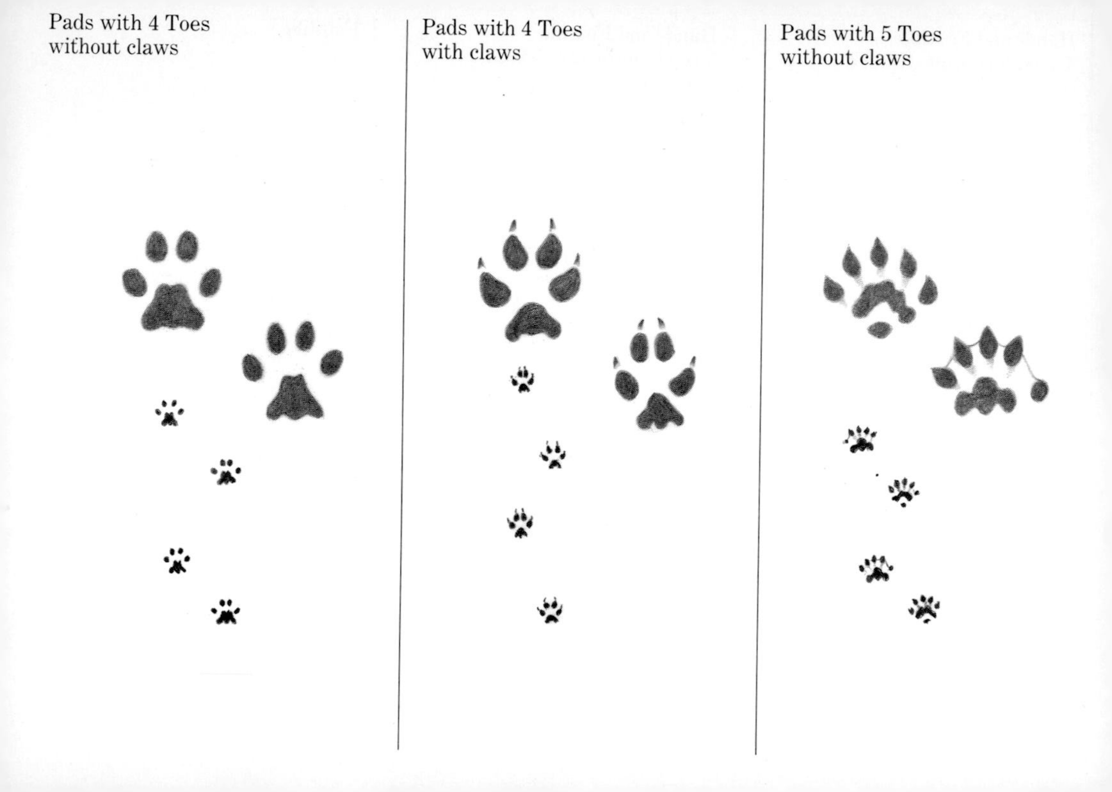
Pads with 4 Toes
without claws
Pads with 4 Toes
with claws
Pads with 5 Toes
without claws

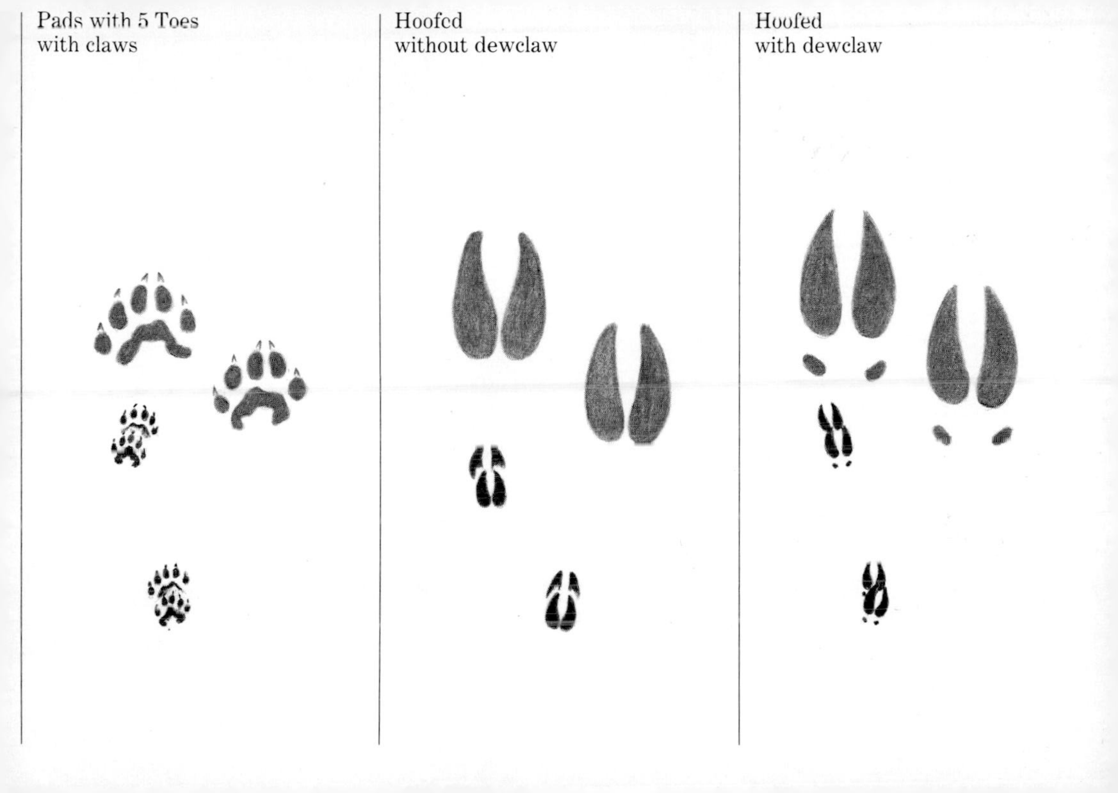
Pads with 5 Toes
with claws
Hoofed
without dewclaw
Hoofed
with dewclaw

The Mammals

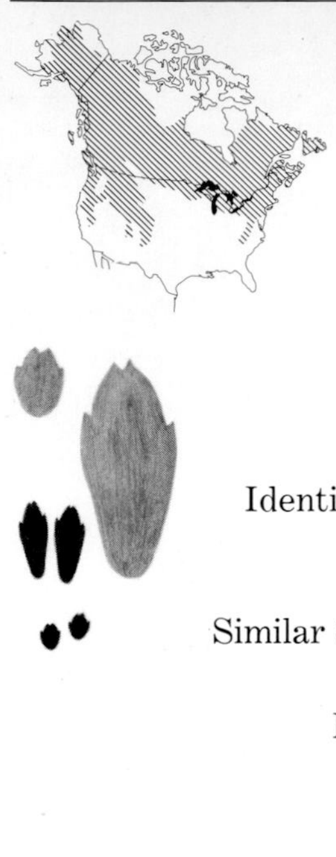

Snowshoe Hare *Lepus americanus*

Perhaps because it is hunted by hawks, owls, foxes, Coyotes, weasels, Wolverines, Lynx, and Bobcats, the Showshoe Hare, or Varying Hare, is shyer than other hares and spends the daylight hours concealed in a shallow bed of grass or a hollow log. Beginning in March, females raise two or three litters a year, each litter containing an average of three young—born with their eyes open, as are all hares. For reasons not fully understood, Snowshoe Hare numbers increase steadily for about 10 years and then undergo a dramatic "crash."

Identification: 17½–21″. A forest-dwelling rabbit with large hind feet. All white in winter, dark brown (including feet) in summer.

Similar Species: Cottontails smaller, brown all year round, feet usually pale. Jackrabbits live in open country.

Habitat: Coniferous forests.

Range: Alaska and Canada south to California, New Mexico, Great Lakes, N. New Jersey, and in Appalachians to Carolinas.

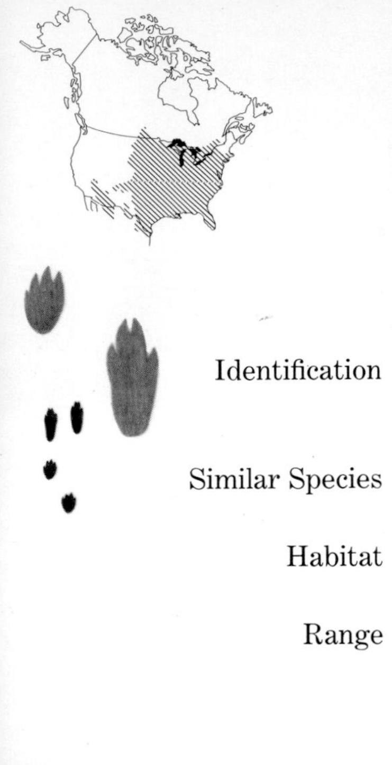

Eastern Cottontail *Sylvilagus floridanus*

This is the most prolific rabbit in eastern North America, raising three or four litters a season and averaging four or five young per litter. Cottontails are fleet-footed and when threatened prefer to hop quickly out of sight into a thicket. If really pressured, they can bound 10 or 15 feet at a time. In places where they are not persecuted, they become quite tame and can often be seen browsing peacefully on lawns. They are most active at night and at twilight.

Identification	14–18″. A small, grayish-brown rabbit with white feet, rusty nape, and often a white spot on forehead. Tail white below.
Similar Species	Marsh Rabbit (*S. palustris*) lacks rusty nape, has dark feet; found in wetlands from SE. Virginia to Florida.
Habitat	Thickets, weedy fields, woodlands, gardens, and cultivated areas.
Range	E. United States west to Great Plains and Texas; also in Arizona.

Black-tailed Jackrabbit *Lepus californicus*

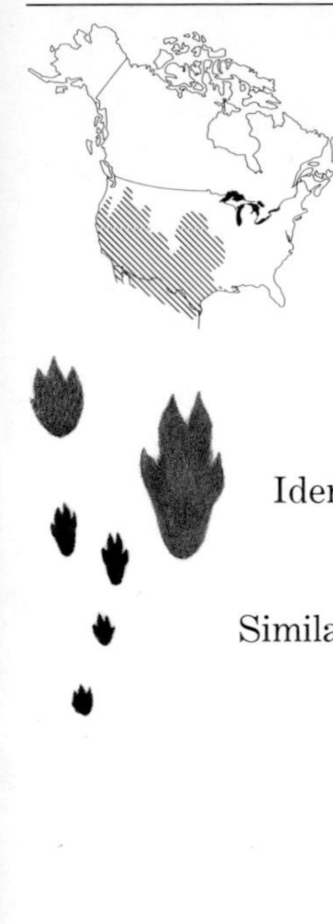

When escaping from a predator, this big, open-country hare can reach speeds up to 35 miles an hour, covering as much as 20 feet in a single leap. At every fourth or fifth leap, a jackrabbit bounds high into the air to take a quick look at its pursuing enemy. Females raise up to four litters a year, usually with two to four young in a litter. By placing each of her young in a separate nest, a mother jackrabbit decreases the chances that she will lose them all to predators.

Identification: 22–24″. A large, gray-brown, long-legged hare with long "jackass" ears that have black tips. Tail has black streak on top.

Similar Species: White-tailed Jackrabbit (*L. townsendii*) has white tail without black streak; found in N. Great Plains and northwestern states.

Habitat: Prairies, meadows, and agricultural areas.

Range: California, central Washington, Idaho, and Nebraska southward into Mexico. Introduced in Kentucky and New Jersey.

Brush Rabbit *Sylvilagus bachmani*

The well-named Brush Rabbit is seldom found very far from thick cover and prefers the densest thickets in chaparral or on forest land that has been cleared. It is most active at night but may be seen foraging quietly on green vegetation during the day. Like other cottontails, it hides its young—which are blind at birth—in a shallow nest covered with grass. Up to five litters, with an average of three young per litter, are raised each year. Young rabbits are more active by day than their parents; if you see a Brush Rabbit during daylight hours, the odds are that it has recently left its nest.

Identification: 12½–14½″. A small, uniformly dark brown, short-eared rabbit with a short, inconspicuous tail.

Similar Species: Desert cottontail (*S. auduboni*) paler with longer ears and legs; found in deserts and grasslands from California east to Great Plains.

Habitat: Dense brushy areas.

Range: Coastal Oregon and California south into Mexico.

Woodchuck *Marmota monax*

A pile of earth in the middle of a meadow, with a large burrow close by, is a sure sign that a Woodchuck is around. Sometimes you will see the burrow's owner standing up inspecting its surroundings; if something alarming shows up, the Woodchuck will give a shrill whistle and dive out of sight. These "groundhogs" have an insatiable appetite for green vegetation—a diet that can make them unpopular with farmers and gardeners. In late summer they start eating even more than usual, putting on fat for the coming months of hibernation.

Identification — 20–27″ (male larger than female). A stocky, grizzled brown or reddish-brown marmot with brushy tail and black feet. Legs short. Ears small.

Similar Species — Yellow-bellied Marmot paler with white spots between eyes; found in mountains south and west of range of Woodchuck.

Habitat — Meadows, fields, roadsides, and woodlands.

Range — Alaska and Canada south to N. Idaho, E. Kansas, E. Oklahoma, Alabama, and Virginia.

Black-tailed Prairie Dog *Cynomys ludovicianus*

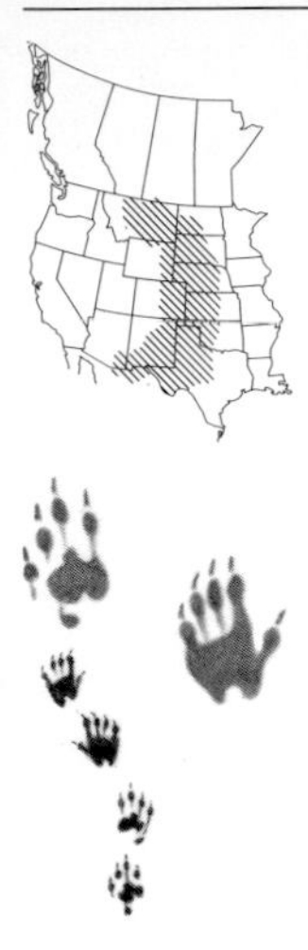

A prairie-dog "town" is a busy and noisy place. Dozens of these sociable animals sit on conical mounds at the entrances to their burrows, barking, whistling, and chirping to one another and keeping a sharp lookout for approaching danger. There is much grooming and "kissing" between relatives, which often have linked burrows and recognize each other as individuals. In early spring, soon after the prairie dogs emerge from hibernation, four or five young are born; they become independent in about 10 weeks.

Identification: 14½–16½". A chunky, yellowish, ground-dwelling squirrel with a slender black tail. Usually seen sitting on a bare mound of earth.

Similar Species: White-tailed Prairie Dog (*C. leucurus*) has white tail, dark smudges above and below eyes; found in open mountain valleys from Montana to New Mexico.

Habitat: Shortgrass prairies and plains.

Range: W. Great Plains from Saskatchewan and North Dakota south to New Mexico and Texas.

Hoary Marmot *Marmota caligata*

One of the characteristic sounds of the mountains of the Northwest is the Hoary Marmot's shrill alarm whistle, delivered from a rocky lookout. Like its eastern relative the Woodchuck, the Hoary Marmot, or "rockchuck," fattens on green plants and hibernates during the colder months. As soon as the snow melts—in February in the southern part of its range and in April farther north—Hoary Marmots reappear and mate. About a month later, four or five young are born naked and blind.

Identification 26–30″ (male larger than female). A large, gray marmot with black on crown and shoulders, black "boots," and reddish tail. Has shrill whistle. Usually heard before seen.

Similar Species Olympic Marmot (*M. olympus*) dull brown without black on crown and shoulders; found only in Olympic Mountains, NW. Washington.

Habitat Rocky slopes in mountains.

Range Alaska and W. Canada south to Washington, Idaho, and W. Montana.

Yellow-bellied Marmot *Marmota flaviventris*

In their mountain habitat, Yellow-bellied Marmots are seen only briefly during the course of a year. They hibernate from early fall until March and then disappear again in June to sleep away the hottest weeks of summer. When they are active, a shrill whistle is often your first clue that a marmot is nearby; then you must carefully scan the rocky slopes around you to find it.

Identification 15–28″ (male larger than female). A yellowish-brown marmot with yellow belly and buff on sides of neck. Head dark, usually with whitish patches between eyes. Feet buff or brown. Has shrill whistle.

Similar Species Hoary Marmot larger and grayer with black feet; found mainly in mountains to north of Yellow-bellied. Woodchuck darker, more uniformly brown.

Habitat Rocky slopes and stony mountain valleys.

Range British Columbia, S. Alberta, and Montana south in mountains to Central California, Utah, Colorado, and N. New Mexico.

Fox Squirrel *Sciurus niger*

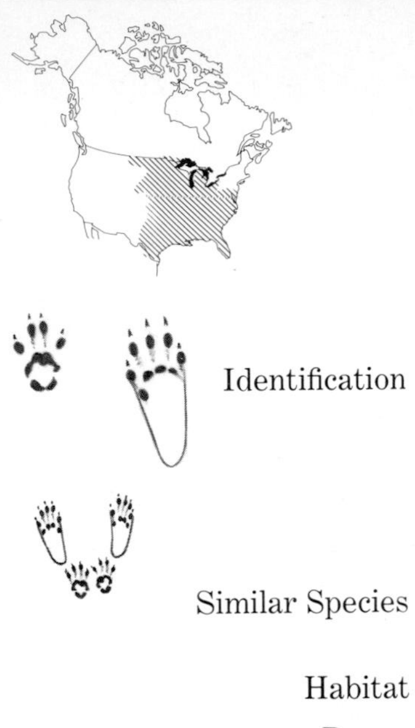

Like the Gray Squirrel, this species eats a wide variety of nuts, fruits, and buds, and its staples are acorns and hickory nuts. In fact, the diet and habits of these squirrels are so similar that they often compete with one another. Surprisingly, the Gray Squirrel seems to win these conflicts of interest. The Fox Squirrel is tamer and, especially in parks and suburbs, is still common throughout the South and Midwest.

Identification 18–27″. A large, variable tree squirrel, usually grayish-buff with yellowish or orange belly and tail bordered tawny. In Southeast often grizzled with black, white or yellow with face marked with black and nose and ears white; some individuals solid black. In Mid-Atlantic states pure gray with no buff or tawny.

Similar Species Gray Squirrel usually smaller and grayer with white-tipped hairs on tail.

Habitat Forests, woodlands, and residential areas.

Range E. United States, except Northeast, west to North Dakota, E. Colorado, and central Texas.

Red Squirrel *Tamiasciurus hudsonicus*

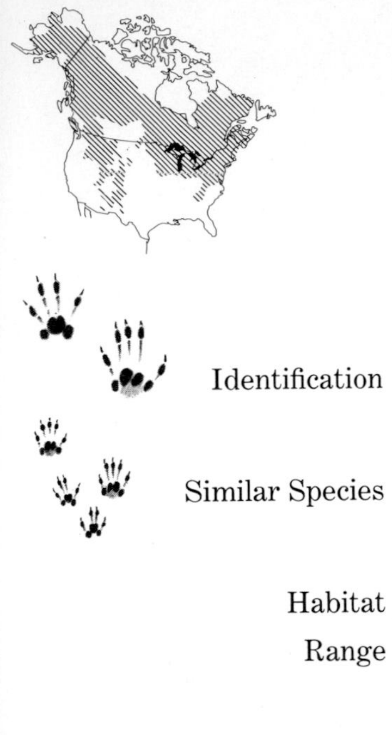

The Red Squirrel is a bundle of nerves, instantly exploding into chatters and wheezy grunts the moment a person or predator appears. It is easy to tell whether there are Red Squirrels in a forest by the little piles of pine-cone fragments they leave behind at favorite eating places. A litter of about four young is born in a woodpecker hole or nest of grass and shredded bark in early spring, and sometimes there is a second litter in late summer.

Identification 12–14″. A small, rust-red or yellowish tree squirrel with whitish belly and bushy tail. In summer usually grayer with black line along side.

Similar Species Douglas' Squirrel (*T. douglasii*) similar but belly grayish or orange; found from SW. British Columbia south to N. California.

Habitat Forests and plantations.

Range Alaska and Canada south to Arizona and New Mexico in Rockies, Great Lakes region, northeastern states, and in Appalachians to North Carolina.

Gray Squirrel *Sciurus carolinensis*

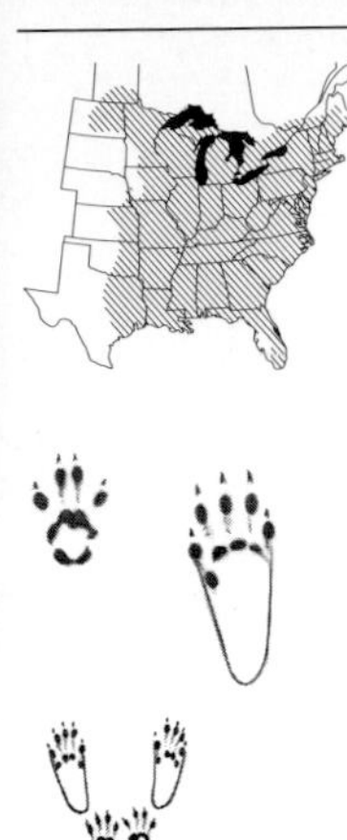

Especially active in early morning and late in the day, Gray Squirrels spend most of their time gathering, storing, or retrieving nuts and seeds. In parks and suburban areas they are very tame and beg insistently for food, but in places where they are hunted they are shy, and you can walk all day through a woodland where many squirrels live without ever seeing one.

Identification 17–20″. A gray tree squirrel with flattened, bushy tail bordered with white-tipped hairs. Belly grayish. In summer tinged with buff on head, upperparts, and feet. In winter back of ears white. Black individuals common in parts of northern states.

Similar Species Typical Fox Squirrel larger with yellow-tipped hairs on tail. Western Gray Squirrel has white belly.

Habitat Forests with nut-bearing trees.

Range E. United States and S. Canada west to Manitoba, Kansas, central Oklahoma, and central Texas. Introduced in Vancouver, British Columbia, and in Seattle, Washington.

Western Gray Squirrel *Sciurus griseus*

This is the only gray tree squirrel in the oak and conifer forests along the Pacific coast. Like its relatives, it eats acorns, nuts, berries, fungi, and pine seeds and builds nests of leaves or twigs during the warm months. In winter it often moves into hollows in trees. Active all year, it raises a litter of three to five young in late winter or early spring and, in the southern part of its range, sometimes a second litter as late as August. The Western Gray Squirrel seems to be more secretive than the eastern Gray Squirrel, but it spends more time foraging on the ground.

Identification: 18–23″. A gray tree squirrel with white belly, hairs in bushy tail a mixture of gray, black, and white, back of ears reddish-brown.

Similar Species: Gray Squirrel (introduced in Washington and British Columbia) has grayish belly.

Habitat: Forests and woodlands.

Range: W. Washington and Oregon south to S. California.

Abert's Squirrel *Sciurus aberti*

Our most boldly patterned squirrel and the only one with long ear tufts, Abert's Squirrel—also called the Tassel-eared Squirrel—is a rather quiet animal that spends all its time in evergreen woodlands and forests. It is seldom as numerous as related squirrels but, perhaps because of its colorful markings, it is easily seen in its open habitat. In spring, three or four young are born in a nest of twigs and needles, and sometimes in the southern part of the range there is a second litter.

Identification — 19–21″. A gray tree squirrel with blackish ear tufts, reddish back, and white belly and underside of tail. Kaibab Squirrel, a form found only on North Rim of Grand Canyon, has blackish belly and all-white tail.

Similar Species — Arizona Gray Squirrel (*S. arizonensis*) gray with white belly, no ear tufts or reddish on back; found only in Arizona and New Mexico.

Habitat — Pine or juniper forests.

Range — Mountains of SE. Utah, Colorado, New Mexico, and Arizona.

Southern Flying Squirrel *Glaucomys volans*

Flying squirrels don't actually fly but use the fold of skin between their front and hind legs to glide from one tree to another. They are nocturnal and spend the day hidden in a woodpecker hole or other tree cavity. To find out if these secretive animals live near you, spread some peanut butter on a tree trunk near a porch light. Any flying squirrels around will soon find it, and you can enjoy watching them from your living room.

Identification 9–9½″. A small tree squirrel with silky fur, grayish-brown above with white belly. Has loose fold of skin between front and hind legs.

Similar Species Northern Flying Squirrel (*G. sabrinus*) larger and browner; ranges from Alaska and N. Canada south to California, Wyoming, Great Lakes region, New England, and in mountains to Carolinas.

Habitat Forests and woodlands.

Range E. United States and S. Canada west to edge of Great Plains; absent from N. New England and S. Florida.

California Ground Squirrel *Spermophilus beecheyi*

Found in open country of many kinds, the California Ground Squirrel lives in colonies. Several animals may share a burrow, but each has its own entrance. Colony sites are used for many generations; the one on the mainland opposite the Seal Rocks in San Francisco has been occupied for years. These ground squirrels hibernate and then mate during a brief period in early spring. A single litter of up to 12 young is born soon thereafter.

Identification 17–19″. A large ground squirrel with bushy, grayish-brown tail edged with white. Brown with buff or whitish flecks, buff belly, vague gray patch on sides of neck, and dark patch on upper back between shoulders.

Similar Species Other ground squirrels in range smaller and lack dark patch between shoulders.

Habitat Open fields, pastures, rocky areas, barren ground, and parks.

Range S. Washington and W. Oregon south through most of California into Mexico.

13-lined Ground Squirrel *Spermophilus tridecemlineatus*

Because of their habit of standing bolt upright to see over the top of the grass, Thirteen-lined Ground Squirrels are also called picket pins. Like most squirrels, these sleek animals are mainly vegetarians, but they also eat insects and birds' eggs and are not above catching a small mammal if the opportunity presents itself. Picket pins are homebodies, hardly ever moving more than a few hundred yards from where they were born.

Identification 8–14″. A tan to brown ground squirrel with 13 whitish lines on back, some broken into rows of whitish spots; belly whitish. No other ground squirrel has pattern of lines and rows of spots on back.

Similar Species Golden-mantled Ground Squirrel and chipmunks more richly colored with 2 or 4 unbroken whitish stripes on back.

Habitat Prairies, fields, roadsides, and lawns.

Range Central North America from Prairie Provinces and Great Lakes region south through Great Plains to Texas.

Golden-mantled Ground Squirrel *Spermophilus lateralis*

This chipmunk-like ground squirrel usually places its burrow near a rock or among the roots of a tree. The tunnel is close to the surface and may be as long as 100 feet. The Golden-mantled Ground Squirrel is common in parks, where it begs for food from campers and hikers. Like its relatives, it has a single litter each year; up to six young are born in early summer.

Identification 10–12″. A small, boldly patterned ground squirrel with coppery-red head and shoulders; brownish, gray, or buff back; and 1 white stripe bordered with black on each side of body. Lacks stripes on face.

Similar Species Cascade Golden-mantled Ground Squirrel (*S. saturatus*) larger with stripes on back more vague; found in Cascade Mountains of British Columbia and Washington. Thirteen-lined Ground Squirrel paler, with pattern of lines and spots on back. Chipmunks smaller with stripes on face as well as back.

Habitat Moist forests and brushy areas.

Range Mountains of W. North America.

Eastern Chipmunk *Tamias striatus*

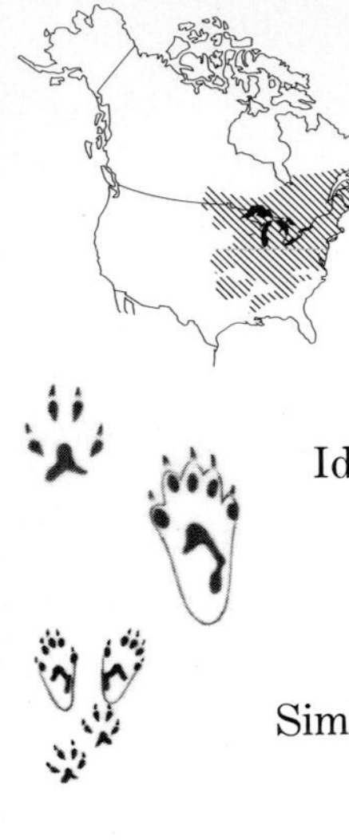

Although they live on the ground and look very much like Golden-mantled Ground Squirrels, chipmunks are not ground squirrels. Instead of living in colonies, they tend to be solitary, and while they may spend the winter sleeping in a burrow, they keep it well stocked with seeds and nuts and do not hibernate as true ground squirrels do.

Identification	9–11″. A small ground-dwelling squirrel with reddish-brown upperparts, white belly, 1 white stripe bordered with black on each side of body, narrow stripes on face, and reddish rump. Only chipmunk in most of E. United States.
Similar Species	Least Chipmunk (overlaps with Eastern in Great Lakes region) smaller with 2 white stripes on each side of body.
Habitat	Forests, woodlands, and residential areas.
Range	Manitoba, Ontario, and Quebec south through most of E. United States to Oklahoma, NE. Louisiana, and Virginia.

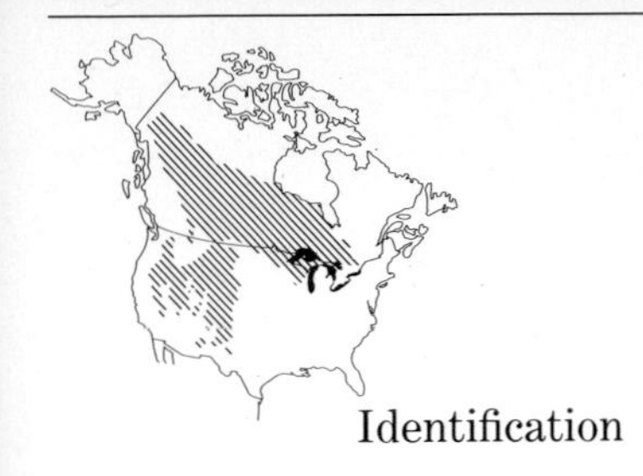

Least Chipmunk *Tamias minimus*

In much of the West this is the chipmunk you are most likely to see. It usually runs with its tail held straight upward. The Least Chipmunk usually lives in a burrow but sometimes nests in a tree. Up to seven young are born in May, and there may be a second litter.

Identification 7–9″. A small and widespread chipmunk. Back yellowish-gray to gray-brown. Two white stripes extending to base of tail, bordered with tan, brown, or black on each side of body. Sides reddish-brown in East, paler in West. Tawny in front of ears.

Similar Species Most other western chipmunks have dark in front of ears. Cliff Chipmunk (*T. dorsalis*) has indistinct stripes on back; found from Idaho south to Arizona and New Mexico. Eastern Chipmunk larger, with 1 white stripe on each side of body.

Habitat Rocky areas, pine woodlands, and sagebrush plains.

Range W. Canada east to Great Lakes region and south through mountains of W. United States; absent from Pacific slope.

Pika *Ochotona princeps*

Whenever you are in the high country, you can be sure you are being watched carefully by several Pikas, each one perched on its favorite rocky vantage point. If you come too close, they will startle you with a short, piercing bleat. Unlike the larger marmots that share their alpine habitat, Pikas do not hibernate. They prepare for winter by harvesting fresh grass, curing it in small piles in the sun, and then storing this “hay” in a den among the rocks. A litter of two to five young—naked and blind like the young of rabbits—is born in May or June, and often a second litter in late summer.

Identification 7–8½″. A small, short-tailed, short-eared relative of rabbits; uniform brown. Has short, bleating call.

Similar Species Collared Pika (*O. collaris*) has gray collar on neck and shoulders; found in Alaska and NW. Canada.

Habitat Rocky mountain slopes.

Range High mountains of W. North America from British Columbia and Alberta south to central California, S. Utah, and N. New Mexico.

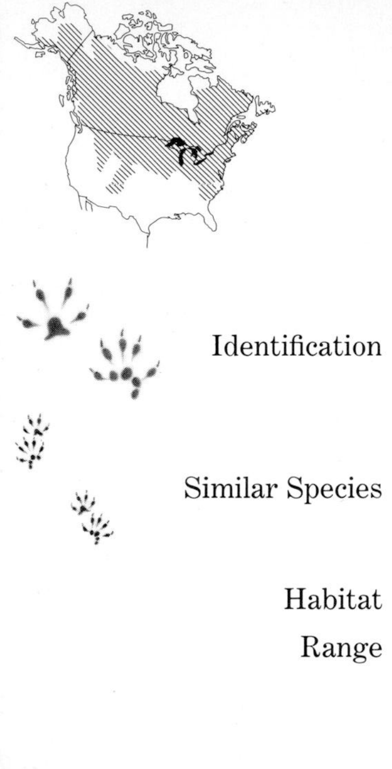

Meadow Vole *Microtus pennsylvanicus*

The abundant Meadow Vole is one of the most important animals in its range because it provides a major food item for owls, hawks, foxes, Bobcats, all of the members of the weasel family, and Raccoons. It is even taken by squirrels and shrews. The vole population can withstand this constant onslaught because of the voles' high birth rate. Young are born in every month of the year; there is a record of a female that had 17 litters in one year.

Identification 5½–8″. A small, variable "field mouse" with long fur varying from dark brown to gray tinged with brown or yellowish grizzled with blackish. Small ears. Tail shorter than body.

Similar Species Southern Red-backed Vole has reddish back, different habitat. Other voles have different ranges or are larger, shorter-tailed, or yellowish on snout.

Habitat Meadows, fields, marshes, and open, grassy woodlands.

Range Alaska, Canada, and N. United States south to Idaho, New Mexico, Missouri, Georgia, and the Carolinas.

Southern Red-backed Vole *Clethrionomys gapperi*

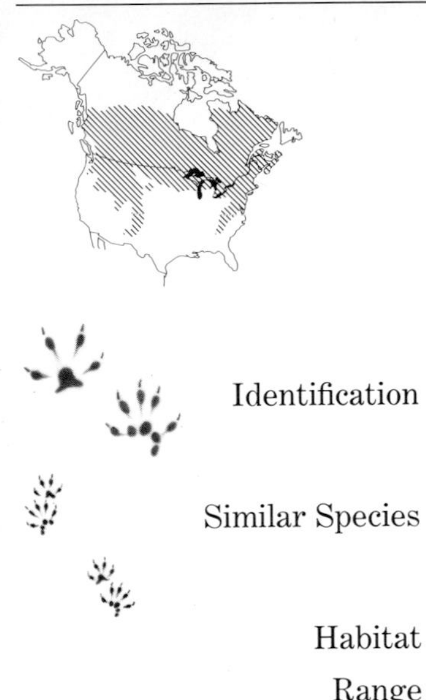

This forest-dwelling vole does not make tunnels like those of the Meadow Vole but uses a system of runways through the grass or lives in the burrows of other animals. Concealed under a log or among the roots of a tree, a female gives birth to several litters of four or five young every year, but a host of predators keep numbers down. These voles eat green plants, seeds, and berries and store more durable items like nuts and tubers.

Identification	5–6″. A small vole with long, soft, brown fur usually tinged reddish in middle of back. Small ears. Tail shorter than body.
Similar Species	Western Red-backed Vole (*C. occidentalis*) similar; found along Pacific coast from British Columbia to N. California. Meadow Vole larger, not reddish.
Habitat	Damp forests, wooded swamps, and bogs.
Range	Canada and N. United States south to Oregon, New Mexico, Arizona, Great Lakes region, Maryland, and in Appalachians to North Carolina.

Meadow Jumping Mouse *Zapus hudsonius*

If you disturb a jumping mouse in its hiding place in tall grass, it will make a few long, fast leaps, drop back out of sight again, and freeze; you are not likely to find it again. But even though these mice are hard to see, it is easy to find out whether they are around. Just look for grass cuttings, made as a jumping mouse bites the stem of a seed head, pulls the stem down, bites it off again, and gradually brings the seeds down within reach. Unlike other mice, jumping mice hibernate.

Identification 8–10″. A long-tailed, long-legged, brown mouse with yellowish tinge on sides, white belly. Best identified by jumping behavior.

Similar Species Western Jumping Mouse (*Z. princeps*) larger; found mainly south and west of Meadow Jumping Mouse.

Habitat Moist meadows, weedy areas, marshes, and brushy woodlands.

Range Alaska and Canada south to Wyoming, Oklahoma, and Georgia.

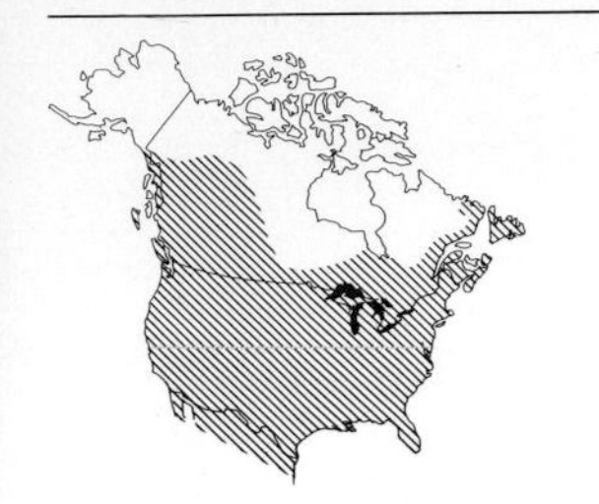

House Mouse *Mus musculus*

The all-too-familiar House Mouse is native to Asia, but long ago it became associated with people and has spread around the world with them. It arrived in North America with the first European explorers and quickly became a major pest. In agricultural areas House Mice often spend the warm months outdoors, but when cold weather comes they move indoors and raid pantries and make nests in sofas and mattresses. House Mice breed very rapidly, especially if they find an abundant food source like a crop of corn. A female is ready to have her first young when only six weeks old and can have as many as 14 litters a year, with up to 12 young per litter.

Identification	6–6½″. A gray or grayish-brown mouse with long, dark, naked tail. Belly gray or buff, almost as dark as back.
Similar Species	Other long-tailed mice are browner, have white belly.
Habitat	Buildings and cultivated areas.
Range	Throughout continental United States and S. Canada.

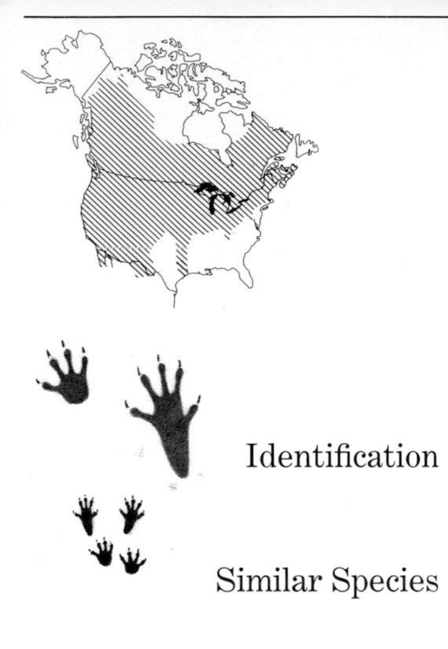

Deer Mouse *Peromyscus maniculatus*

If the House Mouse is the "city mouse," then here is a good candidate for the title of "country mouse." Deer Mice spend most of their lives outdoors where their activities do not interfere with humans. In winter they sometimes seek shelter in warm houses and often bring their stores of nuts and seeds with them. Deer Mice are associated with two serious diseases: Lyme disease, a bacterial infection, and Hanta virus, an often fatal respiratory illness. Avoid contact with Deer Mice and their droppings.

Identification — 6–9″. A long-tailed mouse with grayish or reddish back and sides, white belly and feet. Tail usually dark above and pale below, longer than body.

Similar Species — White-footed Mouse has shorter tail (Northeast) or is not dark above and pale below (South); different habitat.

Habitat — Fields, prairies, and open woodlands.

Range — Canada and continental United States except S. New England, southeastern states, Oklahoma, and Texas.

White-footed Mouse *Peromyscus leucopus*

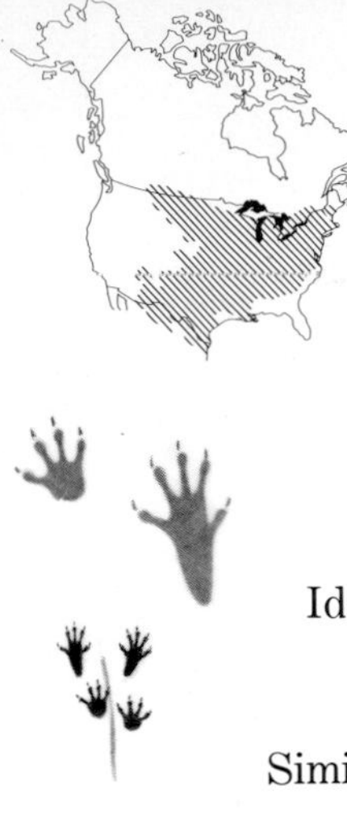

Like its close relative the Deer Mouse, the White-footed Mouse is an occasional winter guest in houses. But just as often, these sleek, nocturnal mice make a warm home by taking over an abandoned bird's nest and roofing it over with twigs and bark strips and lining it with finely shredded grass. Both White-footed and Deer mice are active all winter, and their tracks can often be seen in the snow, especially along the sides of stone walls where they can dart quickly to safety if surprised by an owl, fox, or cat.

Identification	6–8″. A long-tailed mouse with rich, reddish back and sides, white belly and feet. Tail dark both above and below, usually shorter than body.
Similar Species	Deer Mouse often grayer; tail dark above, pale below, and longer than body.
Habitat	Woodlands and thickets.
Range	E. United States and S. Canada west to Montana, Colorado, and Arizona; absent from coastal plain in southeastern states.

Ord's Kangaroo Rat *Dipodomys ordii*

The best way to see a kangaroo rat is to go driving at night in a sandy grassland or desert. Before long, one of these agile rodents will bound through the headlights in front of your car. Kangaroo rats are well adapted for life where water is scarce. They never have to drink because they produce water while digesting seeds; they keep their burrows humid by plugging them up during the day; and they can absorb water vapor from their breath as they exhale, so they avoid losing water in the dry desert air.

Identification 9–11″. A mouselike rodent with long tail, long hind legs, and short front legs. Back and sides buff, belly and feet white. Tail hairy with black and white stripes; black stripes wider than white ones. Best told by distinctive jumping behavior.

Similar Species Other kangaroo rats larger or darker brown above with paler stripes on tail.

Habitat Sandy plains and deserts.

Range Widespread in dry lowlands of W. United States.

Western Harvest Mouse *Reithrodontomys megalotis*

Harvest mice feed almost entirely on the seeds of grasses and weeds, and store them in caches underground. They are active both day and night, foraging on the ground or climbing skillfully in vegetation. Their young—usually two to four to a litter—are often raised in an old woodpecker hole, an abandoned bird's nest, or in a baseball-sized nest of grass built in a shrub or tangle of vines.

Identification 4½–7″. A small mouse with large ears, grayish or brownish back, buff-tinged sides, and white or gray belly. Tail usually dark above, white below.

Similar Species Eastern Harvest Mouse (*R. humulis*) richer brown; found from Ohio and Maryland south to E. Texas and central Florida. House Mouse more uniformly gray; usually found in buildings or croplands.

Habitat Dry grassy or weedy areas.

Range S. British Columbia, S. Alberta, North Dakota, and S. Wisconsin south to NE. Arkansas, W. Texas, and Mexico.

Eastern Woodrat *Neotoma floridana*

Woodrats are also called pack rats or trade rats because they pick up bright, metallic objects such as spoons, belt buckles, coins, cartridges, and even wrist watches and add them to their large, stick nests. They often leave some other object in place of one they have stolen, but this is not due to honesty; when they find a watch or coin, they drop whatever else they are carrying and rush off with their newer discovery.

Identification — 13–16″. A medium-sized, grayish-brown rat with white or pale gray belly, white feet, large eyes and ears, and hairy tail.

Similar Species — Other woodrats are grayer, not tinged with brown, have dusky feet, or have white patch on throat; all found farther west. Norway Rat has smaller eyes and naked, scaly tail.

Habitat — Rocky areas, thickets, and wooded lowlands.

Range — SW. New York, Pennsylvania, and South Dakota south to central Texas and N. Florida.

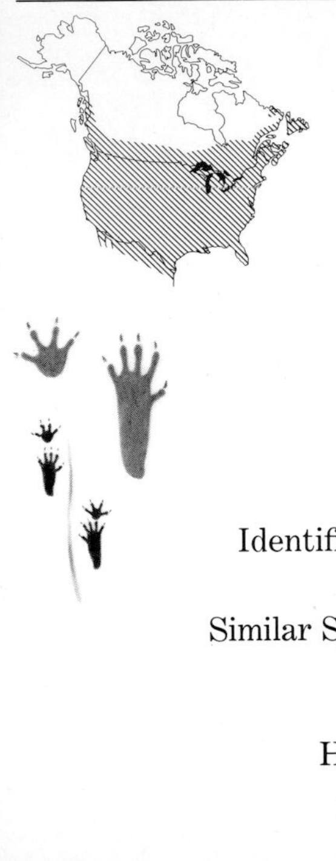

Norway Rat *Rattus norvegicus*

It is hard to find anything good to say about the Norway Rat. Not only do these rats destroy untold tons of food each year, but they are major carriers of diseases, including typhus and bubonic plague. They are so well adjusted to their environment that if their numbers are reduced in a poison campaign, females simply start producing more young, and soon there are just as many rats as there were to begin with. But there is one thing that can be said in favor of this despised mammal. The white laboratory rat, descended from this species, has been used for decades in medical research, and the knowledge from this research has saved countless lives.

Identification 15–16″. A large rat with naked, scaly tail and large ears. Brownish-gray with grayish belly.

Similar Species Black Rat (*R. rattus*) usually darker, has longer tail. Eastern Woodrat and other woodrats white or whitish below with larger eyes and hairy tails.

Habitat Buildings and cultivated areas.

Range Throughout continental United States and S. Canada.

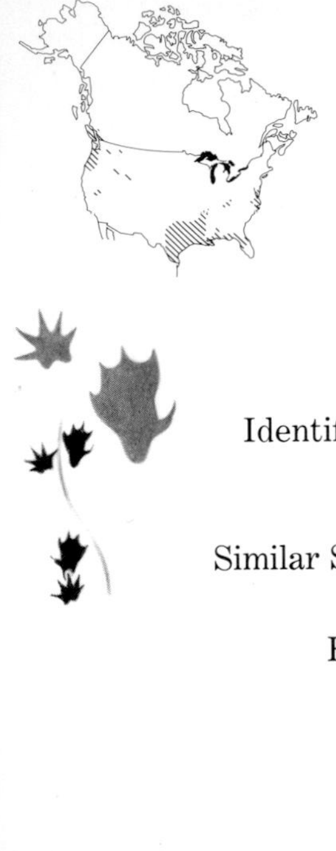

Nutria *Myocastor coypus*

Like many introduced animals, the Nutria is a mixed blessing. It is a valuable furbearer and may control aquatic vegetation that clogs waterways, but it also damages rice and sugarcane crops, digs holes in levees, and competes with the native Muskrat. It is easy to tell if a marsh has Nutrias in it; these robust vegetarians build feeding platforms up to six feet across, make a network of trails through the bulrushes, and at night produce a chorus of piglike grunts.

Identification 26½–55″ (male larger than female). A large, grayish-brown, aquatic rodent with a long, round, scantily haired tail. Hind feet large and webbed.

Similar Species Muskrat smaller with tail flattened from side to side. Beaver much larger with paddle-shaped tail.

Habitat Marshes, streams, and weedy ponds.

Range Introduced from South America. Found in Texas, Louisiana, and eastward along the Gulf Coast; in isolated colonies in New Jersey, Maryland, Great Plains, and Pacific Northwest.

Muskrat *Ondatra zibethicus*

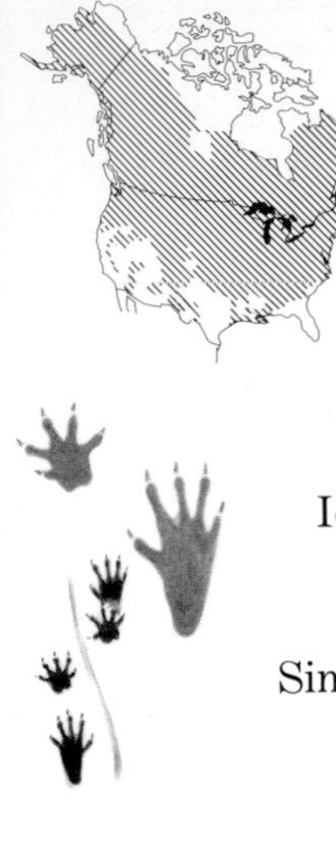

The widespread Muskrat prefers to live in a burrow in the bank of a pond or stream, but in shallow marshes where there are no banks it builds a rounded lodge of cattails or reeds. Wherever Muskrats live, they can be seen swimming slowly through the water, ferrying mouthfuls of plants to their feeding platforms. Active all year, they swim under the ice in winter and often stay submerged for as long as 15 minutes at a time.

Identification	16–24½″. A large, dark brown, aquatic rodent with a long, nearly naked, vertically flattened tail. Hind feet large and partially webbed.
Similar Species	Round-tailed Muskrat (*Neofiber alleni*) smaller with round tail; found in SE. Georgia and Florida. Beaver much larger with paddle-shaped tail. Nutria larger, with tail round, not flattened.
Habitat	Marshes, rivers, and marshy lakes and ponds.
Range	Alaska, Canada, and continental United States; absent from drier parts of Texas and Southwest, S. Georgia, and Florida.

Mountain Beaver *Aplodontia rufa*

Despite its name, the Mountain Beaver is not especially partial to mountains, nor is it a beaver. Primarily a land-dweller, it is a strict vegetarian that lives in an elaborate system of tunnels close to the surface of the ground but with a breeding chamber as much as five feet down. The entrance to a burrow is often covered by a tentlike structure of sticks, leaves, and fern fronds. This rodent is active all year and stores grass and ferns in its runways in late summer and fall. Mountain Beavers raise four or five grayish-brown young in a single spring litter each year.

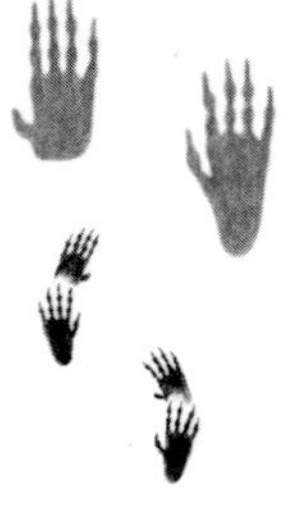

Identification 12½–17″. A small, stocky, dark brown rodent with short legs and very short tail. Not at all like a Beaver.

Similar Species Muskrat larger with long, vertically flattened tail.

Habitat Moist forests and thickets.

Range SW. British Columbia, W. Washington, W. Oregon, NW. California, and in Sierra Nevada of California and W. Nevada.

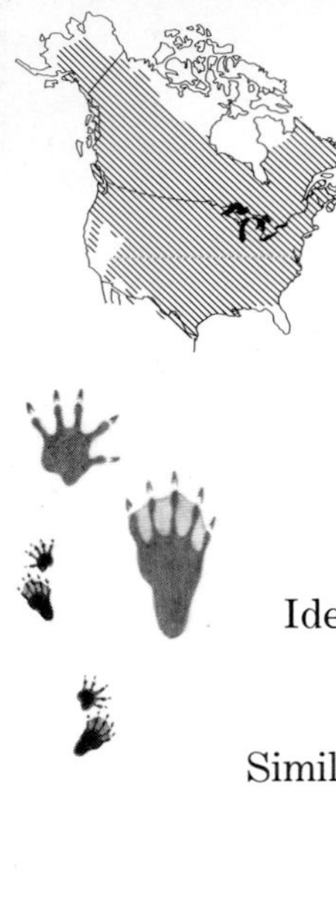

Beaver *Castor canadensis*

Perhaps because so many thousands of Beavers were trapped in the days of the fur trade in both the United States and Canada, many people think this is now a rare animal. But Beavers have been reintroduced in nearly every state, and they are once again common in many areas. They are shy and largely nocturnal, but their presence is easy to detect if you know what to look for. Flooded marshes and meadows, dams and lodges made of sticks, saplings, and mud, tree trunks that have been gnawed or felled, peeled twigs, and the riflelike slap of a tail are signs of these rodents.

Identification 36–47″. A large, dark brown, aquatic rodent with a broad, black, paddle-shaped tail and large, webbed hind feet.

Similar Species Muskrat and Nutria usually smaller with long, slender tails.

Habitat Ponds, marshes, streams, and rivers.

Range Alaska, Canada, and continental United States; absent from most of Florida and dry parts of Southwest.

Porcupine *Erethizon dorsatum*

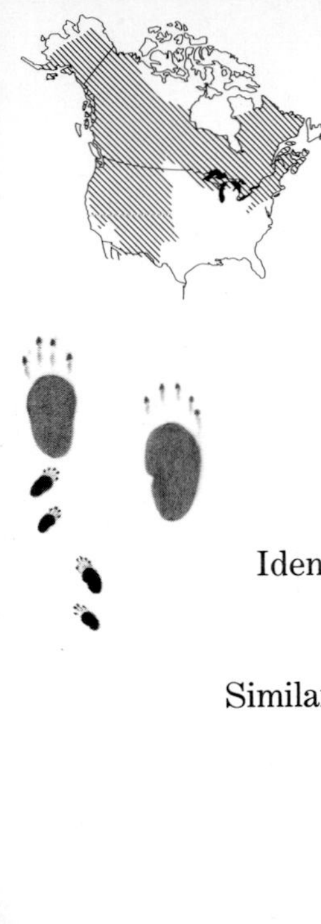

Heavily armed with quills, Porcupines share with skunks the freedom to lead a rather placid existence, safe from any predators that pursue other animals. They feed mainly at night, but you can often find one sitting quietly in a tree and gnawing on bark or ambling along a country road. In parks and other places where they are not molested, Porcupines become tame, and then it is time for you to be careful. Although they cannot throw their quills as many people believe, the quills are barbed and extremely sharp, and the slightest touch can dislodge them from the Porcupine's back or tail and leave them painfully embedded in your finger.

Identification: 26½–40″. A large, stocky rodent with short legs and stout quills on lower back, rump, and tail. Fur long and yellowish in West, blackish or brown in East.

Similar Species: No other mammal has quills.

Habitat: Forests and woodlands, especially with conifers.

Range: Alaska, Canada, W. United States, Great Lakes region, and northeastern states.

Virginia Opossum *Didelphis virginiana*

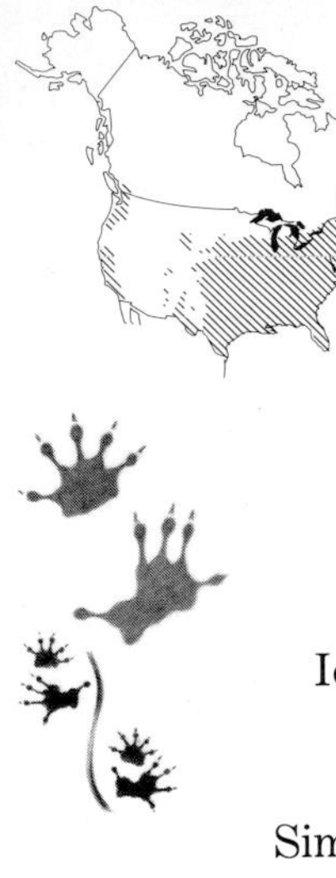

At Jamestown, Virginia, in 1608, Captain John Smith wrote: "An Opassom hath a head like a swine, and a taile like a Rat, and is of the bignesse of a Cat." This is not a bad description of the Virginia Opossum, our only pouched mammal, or marsupial, related to the kangaroos of Australia. Unlike those of any other North American mammal, its young are born after only 13 days and, no bigger than bumble bees, they crawl into the mother's pouch and remain there for about two months. Up to three litters, each containing as many as 12 young, are produced in a year.

Identification: 31–33″. A clumsy animal with long, naked tail; long, pointed snout; large, hairless ears; and long, grizzled, grayish fur. Some animals blackish, especially in South.

Similar Species: Norway Rat and other rats smaller with shorter fur.

Habitat: Woodlands, thickets, farmlands, and residential areas.

Range: E. United States west to Colorado and Arizona; introduced along Pacific coast.

Nine-banded Armadillo *Dasypus novemcinctus*

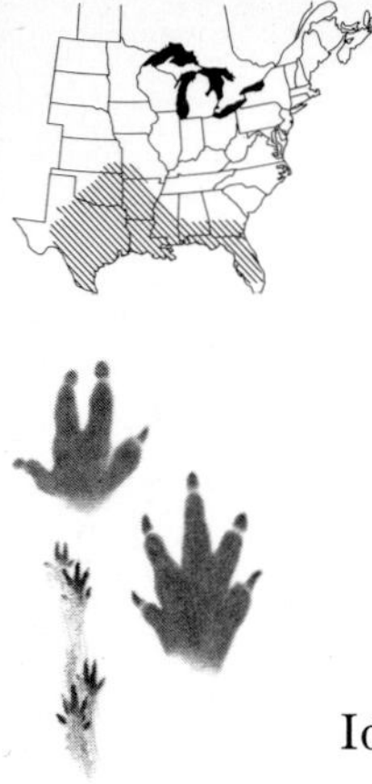

Armadillos are inveterate burrow diggers. They live in burrows, dig in search of insects, and will even try to escape from a predator by quickly excavating a tunnel right on the spot. While this burrowing has made enemies among livestock owners, it is also the secret of this armored animal's success, for armadillos obtain more than half their food underground. Here is one animal that is easy to see because it usually becomes so engrossed in its burrowing that you can walk right up to it. When it finally notices you, an armadillo will scamper away over the leafy ground and disappear.

Identification 31–34″. A short-legged, long-tailed animal with body, top of head, and tail covered with hard, scaly plates. Middle of body has 9 (or fewer) flexible, scaly bands which allow for bending.

Similar Species No other mammal has scaly armor plates.

Habitat Sandy areas and streamside thickets.

Range S. United States from Texas, SE. Kansas, S. Missouri, and Gulf Coast east to Florida.

Star-nosed Mole *Condylura cristata*

This aptly named mole prefers damp soil, although it is sometimes found in dry or hilly country. The "star" on its nose is used to find food in its tunnels, and while the mole is eating, these sensitive projections are folded out of the way. Unlike other moles, it is a good swimmer and even dives and catches minnows. About six young are born in a single litter in the spring and are weaned and independent after only three weeks.

Identification	8″. A black, silky-haired mole with large front feet, a hairy tail, and many fleshy projections around nose. Eyes and ears scarcely visible.
Similar Species	Eastern Mole has short, naked tail, lacks fleshy projections around nose.
Habitat	Moist woodlands and fields, marshes, and residential areas.
Range	SE. Canada and Great Lakes region south to Virginia and in Appalachians to Carolinas; also along coast of Georgia.

Eastern Mole *Scalopus aquaticus*

The bane of anyone who likes a well-tended lawn, this mole makes extensive and conspicuous tunnels just below the surface of the ground. These shallow tunnels are temporary, made as the mole moves along searching for earthworms and burrowing insects. More permanent burrows, including a nest chamber for the four or five young born in the spring, are deeper in the soil.

Identification 5–9″. A silky-haired mole with large front feet and a short, naked tail. Fur varies from gray in North to brownish or tan in South and West. Eyes and ears scarcely visible.

Similar Species Broad-footed Mole (*Scapaneus latimanus*) has short, hairy tail; found in S. Oregon and California. Star-nosed Mole black with fleshy projections around nose. Shrews smaller with more pointed snouts and clearly visible eyes.

Habitat Fields, lawns, and vacant lots.

Range Extreme SE. Wyoming, S. Minnesota, Michigan, and Massachusetts south to Texas and Florida.

Plains Pocket Gopher *Geomys bursarius*

Pocket gophers look like oversized moles, but their large, gnawing front teeth reveal that they are rodents. They usually stay out of sight in their burrows, feeding on roots and tubers. The mounds of soil they bring to the surface make it easy to tell they are there. These mounds are often arranged in a row that shows the direction of the gopher's tunnel.

Identification — 7½–14″ (male larger than female). A thickset, burrowing rodent with prominent front teeth that have 2 grooves; large, strong claws on front feet; small eyes and ears; cheek pouches on outside of face; and a nearly naked tail. Pale brown to yellowish (or almost black in Illinois).

Similar Species — Other pocket gophers in range larger or smaller, or have only 1 groove on each front tooth.

Habitat — Sandy prairies and plains, pastures, and lawns.

Range — Central United States from E. Wyoming, North Dakota, Minnesota, Wisconsin, and Illinois south to New Mexico and Texas.

Masked Shrew *Sorex cinereus*

All shrews are constantly hungry, and none is hungrier than the Masked Shrew, which usually eats its own weight in insects and earthworms every day. It lives its life at a very fast pace, and when startled, its heart rate may rise to more than 1,000 beats a minute. Handling a shrew can be enough to make it go into shock and die, and a single clap of thunder can quite literally frighten a shrew to death.

Identification	3–4½″. A very small, secretive mammal with small ears, pointed snout, and long tail. Fur grayish-brown above, silvery or gray below. Tail dark on top, pale below.
Similar Species	Most other shrews larger, grayer, darker, or with shorter tail. Vagrant Shrew tinged with brown below (summer), grayish or blackish overall (winter).
Habitat	Woodlands, fields, marshes, and bogs.
Range	Alaska and Canada south to Washington, Utah, New Mexico, Kentucky, and Maryland and in Appalachians to Carolinas.

Vagrant Shrew *Sorex vagrans*

A ceaseless hunter of insects, the Vagrant Shrew pursues its prey under leaves and into the runways of voles. When the time comes for breeding, up to nine young are born in a nest of grass concealed in a hollow log or moldering tree stump. Young shrews become independent in about three weeks, and up to then they may follow their mother around in a sort of "train," each one holding onto the tail of the one in front of it.

Identification	4–5″. A small shrew with small ears, pointed snout, and long tail. Fur reddish or brown above and gray tinged with brown below in summer; uniform gray or blackish in winter. Tail usually dark above and below.
Similar Species	Most other shrews larger, grayer, darker, or with shorter tails. Masked Shrew has silvery or gray belly, tail dark on top and pale below.
Habitat	Mixed conifer and broad-leaved forests.
Range	W. Canada south along Pacific coast to central California and in Rockies to Arizona and New Mexico.

Least Shrew *Cryptotis parva*

Most shrews are high-strung, aggressive creatures that are better off living alone, but the Least Shrew is an exception. As many as 30 may live together during the breeding season and in winter, cooperating in maintaining their burrow. But if food becomes scarce, they quickly revert to typical shrew behavior, turning on one another and engaging in cannibalism. The nest of a Least Shrew is a compact ball of grass, hidden under a rock, stump, or even a board in a field. Here a female gives birth to a litter of three to six young, which are weaned in three weeks. Some females have been known to give birth to two litters in a single month.

Identification: 3–3½″. A tiny, warm brown or cinnamon-colored shrew with a short tail. Ears small.

Similar Species: Short-tailed Shrew (*Blarina brevicauda*) larger, lead-gray; found in SE. Canada and NE. United States.

Habitat: Meadows, marshes, and open, moist woodlands.

Range: South Dakota, S. Michigan, New York, and SW. Connecticut south to E. Texas and Florida.

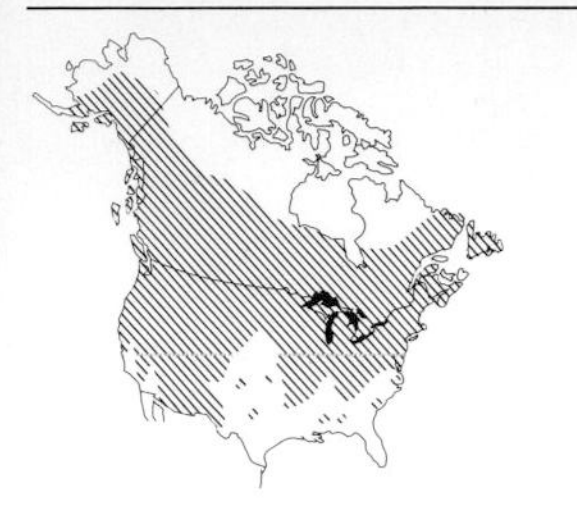

Little Brown Myotis *Myotis lucifugus*

This common bat changes its roosting behavior to suit the season. After mating in the fall, both sexes hibernate in caves or mines. In spring the females gather in warm places like attics to give birth to their single young, and the males, which come out of hibernation later, usually sleep alone under the loose bark of trees. Late in the summer the females leave their warm attics, and both sexes roost together in hollow trees or other dark places.

Identification 3–4″. A small, dark bat, glossy brown above, buff below. Ears short and rounded.

Similar Species Most other similar bats larger, more dull-colored, or with longer ears. Indiana Bat (*M. sodalis*) chestnut-gray with slightly longer tail, smaller feet; found from S. Michigan and New England south to NE. Oklahoma, Missouri, and Alabama.

Habitat Buildings, caves, and mines.

Range Alaska and Canada south to S. California, N. New Mexico, Oklahoma, and Georgia.

Red Bat *Lasiurus borealis*

Unlike most of our other bats, the Red Bat does not gather in roosts but spends the day alone, hanging from a branch or clinging to the trunk of a tree. This and its bright color make it easy to spot and identify. There are other differences, too. It is migratory and in spring and fall often turns up in city parks, far from the usual haunts of other bats. While most bats have only one young at a time, Red Bats often have as many as five, born in late spring. Red Bats commonly hunt for insects around streetlights, and they will even come down to the ground to forage for crickets.

Identification: 4–5″. A small bat with membrane between hind legs furry. Rust-red, frosted with white on back; females duller, paler, and more frosted than males.

Similar Species: Hoary Bat (*L. cinereus*) browner, more frosted; found throughout S. Canada and United States.

Habitat: Forests and woodlands.

Range: S. Canada and United States except Rocky Mountain region and most of Florida.

Brazilian Free-tailed Bat *Tadarida brasiliensis*

Most bats gather in communal roosts, but no roosts are larger than those of the Brazilian Free-tailed Bat. Nearly every town in the southern United States has at least one old mansion or warehouse with a colony of thousands of these highly social bats, and caves in the Southwest, like Carlsbad in New Mexico, may have roosting populations that run into the millions. The evening exodus of these cave-dwelling bats is a spectacular sight, and the tons of insects they consume each year make them a valuable ally of the farmer.

Identification: 3½–4½″. A small bat with tail extending well beyond edge of membrane between hind legs. Fur dark brown or dark gray.

Similar Species: Other free-tailed bats larger and found only in Southwest or S. Florida.

Habitat: Buildings, caves, and mines.

Range: Oregon, Utah, Nebraska, Louisiana, and South Carolina south to Mexico and Gulf Coast; absent from S. Florida.

Arctic Fox *Alopex lagopus*

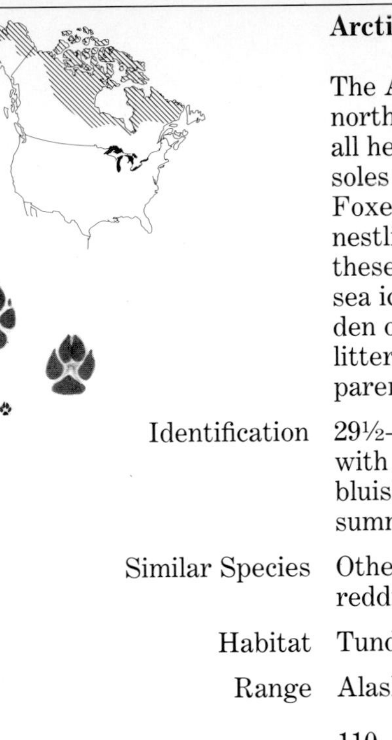

The Arctic Fox has many adaptations for life in its bleak northern habitat. Its thick fur, short legs, and short ears all help conserve heat, and for added protection, the soles of its feet are covered with fur. In summer, Arctic Foxes eat almost anything—small rodents, the eggs and nestlings of birds, and berries—and during the winter, these resourceful predators follow Polar Bears onto the sea ice to scavenge at their kills. In spring, choosing a den on a protected slope, a female gives birth to a single litter of 1 to 14 young which are cared for by both parents.

Identification: 29½–36″ (male larger than female). A small, stocky fox with short legs; short, rounded ears; and bushy tail. Dull bluish-brown or gray with white belly and sides in summer; white or pale slate gray in winter.

Similar Species: Other foxes larger with pointed ears. Red Fox usually reddish, has white tail tip.

Habitat: Tundra and sea ice.

Range: Alaska and N. Canada.

Red Fox *Vulpes vulpes*

Despite centuries of hunting, Red Foxes are still common even on the outskirts of our largest cities. Shy and nocturnal, they are adept at keeping out of sight. Often the best clue that foxes are around is their calling during the winter mating season; vixens have a throaty bark and males a shrill yapping. As many as 10 young are born in a den in early spring and stay with their parents until late summer or fall. There is evidence that Red Foxes mate for life.

Identification 35½–45″ (male larger than female). A large fox with pointed ears. Usually reddish-yellow with white belly and throat, black legs and feet. Some almost solid black, others black frosted with white or brown with dark "cross" on shoulders. All have white tail tip.

Similar Species Other foxes lack white tail tip.

Habitat Forests, thickets, brushy fields, and agricultural areas.

Range Alaska, Canada, and continental United States except Pacific coast, Southwest, Great Plains, coast of Carolinas, and Florida.

Gray Fox *Urocyon cinereoargenteus*

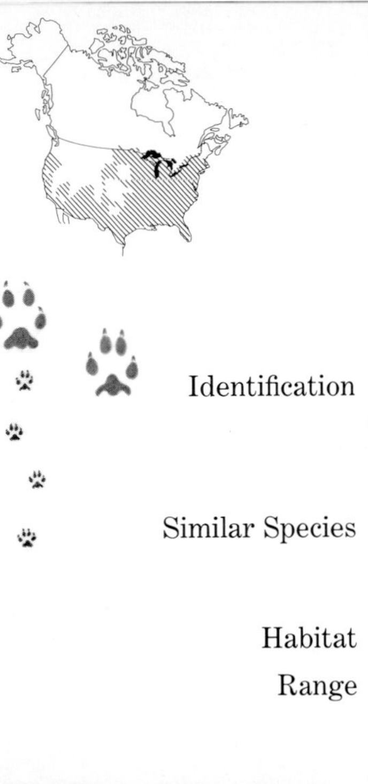

Less often seen than the Red Fox, the Gray Fox is a shyer and quieter animal and prefers denser cover. But like many secretive animals it is also curious, and you can sometimes attract one into view by making a soft, squeaking noise. Gray Foxes seldom dig their own dens; they use hollow logs and natural cavities in rocks or enlarge the abandoned burrows of other animals. They are skillful at climbing trees to feed on birds' eggs or to escape a pack of pursuing hounds.

Identification 31–44″. A large fox with pointed ears. Salt-and-pepper gray above, reddish below and on sides of neck, and white on throat. Tail has "mane" of black hair down middle and black tip.

Similar Species Red Fox has white tail tip. Channel Islands Gray Fox (*U. littoralis*) smaller; found on islands off S. California. Other foxes smaller.

Habitat Forests, woodlands, and thickets.

Range Oregon, Great Lakes region, S. Quebec, and Maine south to Mexico, Gulf Coast, and Florida.

Kit Fox *Vulpes macrotis*

This small fox of open country is not as shy as other foxes and is easily seen if you go driving in the desert at night; sooner or later, a Kit Fox will dart across the road into the beams of your headlights. Kit Foxes dig extensive burrows in soft or sandy soil and sometimes move into an abandoned Badger's den or take over the burrow of a prairie dog. Here they raise a litter of four to seven young which are weaned after about 10 weeks. The family stays together from early spring until fall.

Identification 26–34″. A small, slender fox with large, pointed ears. Pale gray tinged with buff or yellowish. Tail has dark tip.

Similar Species Swift Fox (*V. velox*) slightly larger with shorter ears; found in Prairie Provinces and Great Plains states. Other foxes larger.

Habitat Dry, shortgrass prairies, sagebrush plains, and deserts.

Range Central California, S. Oregon, Utah, SW. Colorado, and W. Texas south to Mexico.

Coyote *Canis latrans*

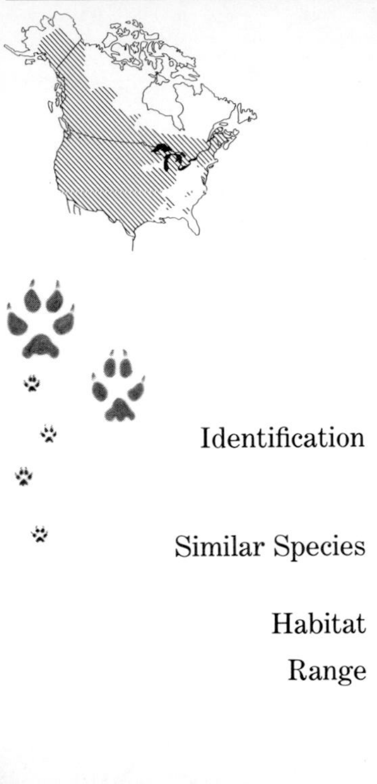

In its ancestral habitat in western grasslands and deserts, the Coyote preyed on small mammals, reptiles, insects, and fruit, and left the bigger game to wolves. Now that wolves have all but disappeared from the United States, Coyotes have expanded their range and are found in most of the country. In some areas, by hunting in packs like their larger cousins, they can bring down a deer—something they could never have done when they hunted alone or in pairs on the plains. For some reason, the well-known howling of the Coyote is seldom heard outside its original western home.

Identification: 41½–52″ (male larger than female). A gray or reddish-gray wild dog with buff belly and legs and dark tail tip. Carries tail pointed downward when running.

Similar Species: Gray Wolf larger, carries tail straight out behind when running.

Habitat: Plains, prairies, woodlands, and brushy areas.

Range: Alaska, W. and S. Canada, and most of continental United States; isolated areas in Southeast.

Gray Wolf *Canis lupus*

Hunted and poisoned, the Gray Wolf has retreated into Canada and Alaska and is now rare in the continental United States. A wolf pack usually numbers up to 15 animals, led by a strong male and consisting of a family group and its relatives. Members of a pack cooperate in hunting in a territory that may cover as much as 260 square miles. They prey on deer, Caribou, and Moose.

Identification: 3½–6½′ (male larger than female). A large wild dog with long, bushy, black-tipped tail. Usually gray grizzled with black but ranges from black to almost pure white. Carries tail straight out when running.

Similar Species: Red Wolf (*C. rufus*) smaller, reddish-brown grizzled with black or all black; found in Texas and Louisiana, but almost extinct. Coyote smaller, carries tail pointed downward when running.

Habitat: Forests and tundra.

Range: Alaska and Canada south to N. Washington, N. Idaho, N. Montana, and Great Lakes region; formerly much more widespread.

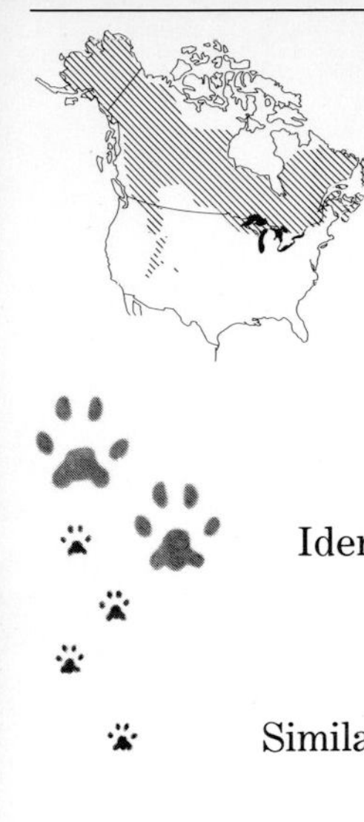

Lynx *Felis lynx*

A silent hunter in dense forests, the Lynx preys mainly on the Snowshoe Hare, which it stalks for miles through the snow or surprises in a sudden pounce from an overhanging branch. Although this big cat will take other prey, it is so dependent on the Snowshoe Hare that when the hare population undergoes one of its "crashes" every 10 years or so, the number of Lynx also drops. The Lynx makes its den in hollow logs or under the roots of a fallen tree.

Identification 29–42″ (male larger than female). A long-legged, large-footed cat with long, black-streaked "sideburns"; long, black ear tufts; and a stubby tail tipped black. General color buff or sandy with scattered black hairs.

Similar Species Bobcat has vague spots, shorter legs, ear tufts, and "sideburns"; tail not completely tipped black.

Habitat Heavy coniferous forests.

Range Alaska and Canada south to Washington, Colorado, the Great Lakes region, and N. New England.

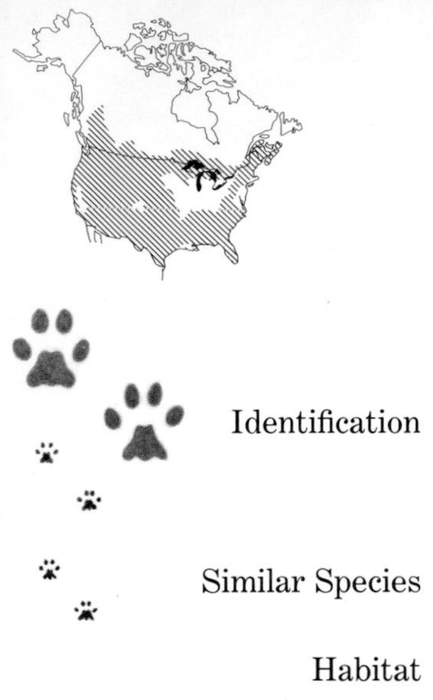

Bobcat *Felis rufus*

A southern relative of the Lynx, the Bobcat is found in a wider variety of habitats and takes a greater variety of prey. Hunting alone and mainly at night, it hides in a thicket, hollow log, or rocky crevice during the day and can go undetected in areas where it is fairly common. Even when it is in full view, its mottled coat makes it easy to overlook. The Bobcat's favorite prey is rabbits, but it also takes rats, mice, squirrels, other mammals, birds, and even frogs.

Identification 28–49½″ (male larger than female). A large-footed cat with short, black-streaked "sideburns"; short ear tufts; and a stubby tail, black on top but not tipped black. General color buff or sandy with vague black spots.

Similar Species Lynx has longer legs, ear tufts, and "sideburns"; lacks vague spots; has tail tipped black.

Habitat Brushy or rocky country, forests, woodlands, swamps, farmlands, and deserts.

Range S. Canada and continental United States; absent from parts of Midwest.

Mountain Lion *Felis concolor*

The Mountain Lion or Cougar is one of North America's best known mammals, but it is rare and elusive and few people have ever seen it. Because of its size, the Mountain Lion preys mainly on large mammals, especially deer, but supplements its diet with insects and other small animals. It is a solitary hunter, living alone except when males and females come together during the mating season. In summer a female produces a litter of up to six spotted kittens and remains with them for at least a year.

Identification	6½–8′. A very large sandy or grayish cat with a long, dark-tipped tail and a dark patch on each side of muzzle.
Similar Species	No other native cat is as large or has as long a tail.
Habitat	Mountain forests, desert hills, and swamps.
Range	British Columbia and S. Alberta south to California and Texas; also in Louisiana, Tennessee, S. Florida, and in scattered localities elsewhere in East. Formerly much more numerous and widespread.

Harbor Seal *Phoca vitulina*

The only common seal along the coasts of the United States, the Harbor Seal can often be found basking in groups of 100 or more on an undisturbed sandbar or ledge when the tide is out. At high tide the seals pursue fish and squid, often entering harbors and estuaries. The single pup is usually born in spring and lacks the woolly white coat of newborn arctic seals.

Identification 4–6′. A sleek seal with a short, doglike muzzle; large eyes; hind flippers that cannot be turned forward; and no external ears. Usually dark yellowish-gray or tan with irregular dark spots and markings that may cover most of body; sometimes all gray or blackish.

Similar Species California Sea Lion has external ears, no spots, and hind flippers that can be turned forward. Gray Seal (*Halichoerus grypus*) larger with longer, horselike muzzle; found from Labrador south to N. New England.

Habitat Sea coasts and harbors.

Range Pacific and Atlantic coasts from Arctic south to S. California and Carolinas.

California Sea Lion *Zalophus californianus*

Familiar as the ball-balancing "seal" of the circus, the California Sea Lion breeds in noisy island rookeries off the coast of southern California. Here pups are born in spring or early summer; each pup stays with its mother for nearly a year. Foraging mainly at night, adults feed on fish, squid, and abalone, but a mature bull standing guard over his territory does not eat at all.

Identification 6½–8′ (male); 4–6½′ (female). An agile sea lion with external ears and hind flippers that turn forward. Uniform tan or brown (looks blackish when wet). Barks frequently when resting on rocks. Male has thick neck and high forehead; female sleeker.

Similar Species Northern Sea Lion (*Eumetopias jubatus*) larger, paler, and usually silent; found from Alaska south to S. California. Harbor Seal spotted; no external ears; hind flippers cannot be turned forward.

Habitat Rocky and sandy coasts and offshore islands.

Range Pacific coast from British Columbia south into Mexico.

River Otter *Lutra canadensis*

During most of the year, otters are among the most playful animals. They seem to make a game out of everything—swimming, diving, and even pursuing prey through the water. The fun stops briefly in spring when females leave the group and give birth to a litter of two to five young, and the males fight for the attention of the females.

Identification	35–52″ (male larger than female). A long, sleek, aquatic mammal with webbed feet and long, tapering tail. Dark brown (looks blackish when wet) with silvery throat and white whiskers.
Similar Species	Sea Otter (*Enhydra lutris*) larger with grayish head; lives in salt water; found along Pacific coast from Alaska to California.
Habitat	Rivers, streams, lakes, ponds, and marshes.
Range	Alaska and Canada south to N. California, Utah, Great Lakes region, and New Jersey; also from Maryland south to Florida and west to Texas. Formerly in Midwestern states.

Long-tailed Weasel *Mustela frenata*

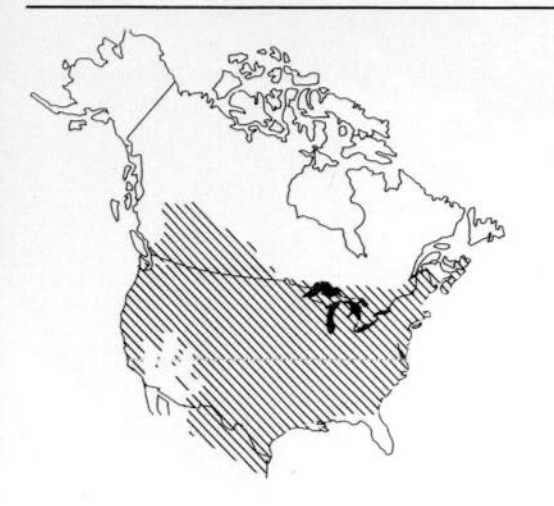

Supple and steamlined, weasels are expert swimmers, skilled tree-climbers, and can follow their prey into the narrowest tunnels and crevices. They hunt small mammals, birds, frogs, and snakes, quickly dispatching their victims with a bite at the back of the neck. They are mainly nocturnal, but now and then they hunt during the day.

Identification: 11–18″ (male larger than female). A slender predator with short, rounded ears; short legs; and long, black-tipped tail. Brown above and on legs and feet; white below, often with yellowish tinge on throat and breast. In North in winter, fur of some animals all white except for tail tip.

Similar Species: Ermine (*M. erminea*) smaller with shorter tail and white feet in summer; found from Alaska and Canada south to California, New Mexico, Michigan, and Maryland.

Habitat: Woodlands, fields, brushy areas, and farmlands.

Range: S. Canada and continental United States except SE. California and desert regions of Arizona.

Fisher *Martes pennanti*

Contrary to its name, the Fisher does not catch fish. But few other creatures are safe from this voracious hunter; a roster of its prey animals reads like a list of the mammals and birds of the northern forests. Even Porcupines are no match for a Fisher, which simply rolls one of these slow-moving mammals over and attacks its unprotected belly. Once nearly wiped out by the fur trade, Fishers are making a slow comeback in parts of New England and elsewhere.

Identification: 31–42″ (male larger than female). A long, thin predator with small ears and a bushy tail. Dark brown or blackish, frosted with white on head.

Similar Species: Mink smaller with white patch on chin. Marten (*M. americana*) smaller with buff breast; found in same range as Fisher.

Habitat: Forests.

Range: Alaska, Canada, Great Lakes region, N. New York, and N. New England; also in scattered areas in mountains of West and in Appalachians.

Mink *Mustela vison*

Unlike River Otters, Minks are anything but sociable. Every Mink vigorously defends a territory of its own, and only in the breeding season can it tolerate other Minks without fighting. Males breed with several females but finally choose a single mate. A frequently used den site is a Muskrat lodge, which may be vacant because its rightful owner has been eaten by the new occupants. As many as six young are born in a fur-lined nest in spring. They stay with their mother until fall, when all Minks return to a solitary way of life.

Identification: 19–28″ (male larger than female). A sleek, bushy-tailed predator. Dark, lustrous brown or blackish with white patch on chin. Usually seen bounding along edge of water.

Similar Species: Fisher larger, lacks white chin patch. Long-tailed Weasel paler brown with white underparts.

Habitat: Shores of rivers, streams, lakes, ponds, and marshes.

Range: Alaska, Canada, and continental United States; absent from desert Southwest and from Newfoundland.

Striped Skunk *Mephitis mephitis*

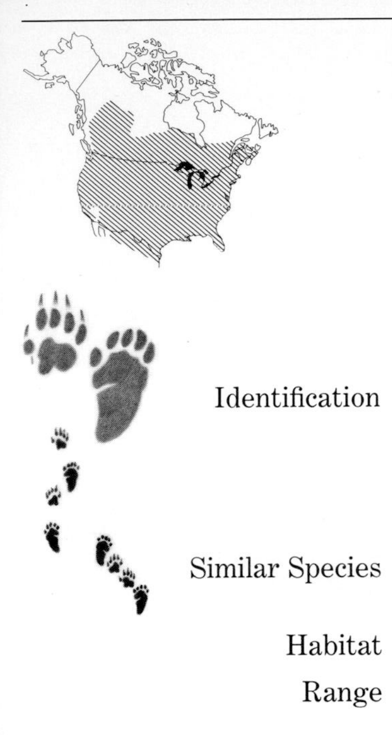

Skunks are docile animals and usually seem reluctant to use their powerful, defensive spray. Instead, they first act out a series of warnings. These include arching the back, stamping the front feet, clicking the teeth, shuffling backwards, raising the tail, and snarling softly. If all of these warnings are ignored, a jet of pungent, sulphur-containing alcohol, which can cause intense pain and temporary blindness, is fired a distance of at least 10 feet.

Identification 22–32″ (male larger than female in most areas). A stocky, short-legged mammal with bushy tail. Black with 2 broad white stripes on back, meeting on top of head and shoulders; a thin white line down forehead; and often with white in tail.

Similar Species Western Spotted Skunk smaller, black with white spots and stripes.

Habitat Woodlands, plains, deserts, and residential areas.

Range S. Canada and continental United States, except parts of desert Southwest.

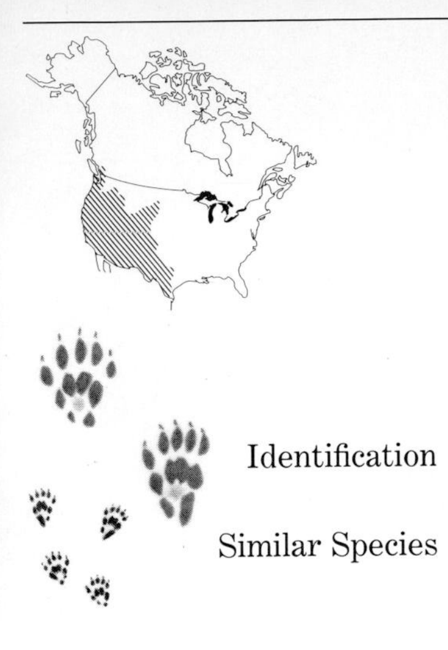

Western Spotted Skunk *Spilogale gracilis*

The warning behavior of a Western Spotted Skunk includes a striking handstand, with the head and contrastingly striped and spotted back pointed toward the enemy. This posture is so convincing that these agile little skunks seldom have to back up their threat by spraying. Although like other members of the weasel family they prey on small mammals, insects, and birds' eggs, Western Spotted Skunks also eat a variety of fruits and are fond of ripe corn.

Identification	13½–23″. A small skunk, black with complex pattern of spots and stripes and usually with white tail tip.
Similar Species	Eastern Spotted Skunk (*S. putorius*) very similar; found in central and SE. United States. Striped Skunk larger with 2 broad stripes along back.
Habitat	Deserts, grasslands, brushy areas, and woodlands.
Range	SW. British Columbia, Idaho, and SW. North Dakota south to Mexico and east to Wyoming, central Colorado, and W. Texas; absent from parts of S. California.

Ringtail *Bassariscus astutus*

The secretive Ringtail hunts mainly at night, often lying in wait for a passing mouse or rabbit. But its diet is much more varied than that and includes birds, snakes, lizards, toads, insects, spiders, scorpions, and fruit. The Ringtail's sharp claws make it an expert climber, and its body is slender enough to enable it to crawl down narrow passageways in pursuit of its prey. Up to four young are born in spring and are tended by both parents; they are white at birth but soon acquire the brown fur and banded tail of the adults.

Identification 24–32″ (male larger than female). A slender, short-legged predator with large eyes and ears, pointed muzzle, and long, bushy tail. Sandy brown with white rings around eyes and blackish bands on whitish tail.

Similar Species Raccoon larger and stockier with black mask; tail shorter with few bands.

Habitat Rocky slopes and deserts; woodlands.

Range SW. Oregon, Utah, Colorado, and S. Kansas south to Mexico.

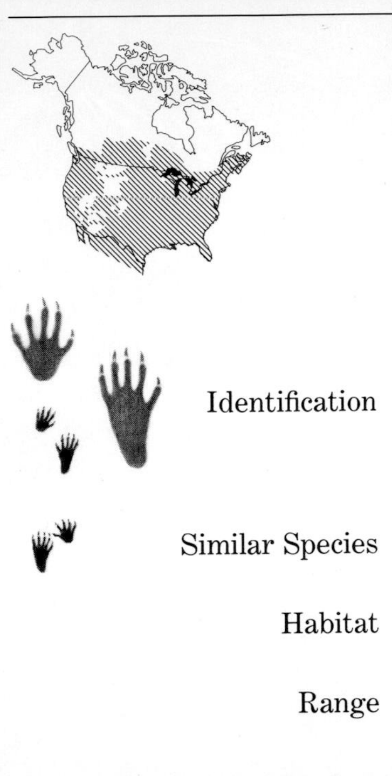

Raccoon *Procyon lotor*

One of North America's most versatile and adaptable animals, the Raccoon is an expert climber and skilled swimmer whose deft fingers can take the lids off garbage cans, turn doorknobs, and locate prey in small crevices. Raccoons are primarily nocturnal and may seldom be seen. They often den in hollow trees, where four or five young are born in April or May. About eight weeks later, the mother leads her family down to the ground and begins to teach them how to forage on their own.

Identification — 26–40″. A chunky, bushy-tailed predator with pointed muzzle. Reddish-brown or grayish-brown with black mask bordered white and with alternating blackish and white bands on tail.

Similar Species — Ringtail smaller and more slender with more dark bands on long tail and no black mask.

Habitat — Forests, woodlands, and residential areas, especially near water.

Range — S. Canada and continental United States; absent from parts of Rocky Mountain states.

Badger *Taxidea taxus*

At first sight, Badgers seem clumsy and not well adapted for capturing rodents. But they are champion diggers and obtain most of their prey by excavating burrows. They also use their digging skills to avoid danger. Since they cannot run very fast, they hastily dig a burrow, wheel around, and face their opponent with sharp teeth and claws. Sometimes they dig so fast they simply disappear into the ground.

Identification 22–30″. A stocky, broad-shouldered, short-legged predator with a short, bushy tail. Fur shaggy, grizzled gray and yellowish-buff with white stripe from upper back to tip of dark snout, black-and-white cheek patches, and black feet.

Similar Species Unmistakable.

Habitat Plains, prairies, and farmlands.

Range British Columbia, Prairie Provinces, and Great Lakes region south throughout W. United States; absent from Pacific coast of Washington, Oregon, and N. California.

Wolverine *Gulo gulo*

For its size, the Wolverine is the most ferocious mammal in North America. It is primarily a carrion-eater but armed with stong jaws and bone-crushing teeth, it preys on a variety of mammals and birds and is capable of bringing down a weakened or snowbound deer, Elk, or Moose. Persistent and fearless, it will drive bears and Mountain Lions away from a kill. The hunting range of a Wolverine may cover as much as 1,000 square miles of northern wilderness, which it patrols with a steady and tireless loping gait.

Identification: 37–41″ (male larger than female). A large, stocky, short-legged predator with small ears and short, bushy tail. Dark brown with pale bands from shoulders back to hips and meeting over tail. Large pale patch in front of ears.

Similar Species: Unmistakable.

Habitat: Forests and tundra.

Range: Alaska and Canada south in mountains to California and Colorado; formerly in northeastern states.

Black Bear *Ursus americanus*

With a diet as much as 95 percent vegetarian, Black Bears seldom prey on anything larger than a squirrel. When fall comes, they gorge themselves on berries, nuts, and roots and prepare for a winter's slumber, but not the deep sleep of true hibernation. Every second January, females give birth to two cubs, which emerge from the den in spring with their mother and usually remain with her for over a year.

Identification 5–6′ (male larger than female). A medium-sized bear with long facial profile. Fur usually black but often cinnamon-brown in West and sometimes white on coast and islands of central British Columbia. Muzzle tan.

Similar Species Grizzly Bear larger with flatter facial profile.

Habitat Forests, swamps, and wooded mountain slopes.

Range Alaska and Canada south in western mountains to Mexico and to Great Lakes region, New York, New England, and in Appalachians to Georgia; also in Florida, Louisiana, and Mississippi Valley. Formerly more widespread and now increasing again.

Grizzly Bear *Ursus arctos*

The Grizzly, or Brown, Bear was once a familiar animal throughout the West. But after more than a century and a half of persecution south of the Canadian border, it survives in just a few wilderness areas in the Rockies. Even here this potentially dangerous carnivore comes into occasional conflict with humans. Only in Alaska and Arctic Canada can one count on seeing Grizzly Bears living as they always have, foraging for berries and mice, leading their cubs across the open country, and gathering to fish when the salmon are running.

Identification	6–8′ (male larger than female). A very large, bulky bear with flat facial profile and distinct hump on shoulders. Fur usually brown but varies from tan to blackish and often grizzled with white.
Similar Species	Black Bear smaller with more pointed facial profile.
Habitat	Open grassy areas in mountains; also in lowlands in North.
Range	Alaska and W. Canada south in mountains to Wyoming.

Polar Bear *Ursus maritimus*

Perhaps because of its white camouflage, the Polar Bear hunts in broad daylight, unlike other North American bears. While both the Grizzly and Black bears are mainly vegetarians, the Polar Bear is a confirmed flesh-eater, preying on seals, Walrus, fish, and birds. In their cold northern habitat, females move into a winter den in November and do not reappear until the following March, when they are accompanied by one to four cubs. The males are more active and spend less time in winter dens. Except during the brief spring mating season, the sexes go their separate ways, perhaps because a male will prey on cubs not guarded by their mother.

Identification	7–11′ (male larger than female). A large, long-legged, long-necked bear with small ears. Fur long, thick, and whitish.
Similar Species	Unmistakable.
Habitat	Arctic coasts, sea ice, and tundra.
Range	Coast and islands of Arctic Ocean and Hudson Bay from Alaska east to Newfoundland.

Collared Peccary *Tayassu tajacu*

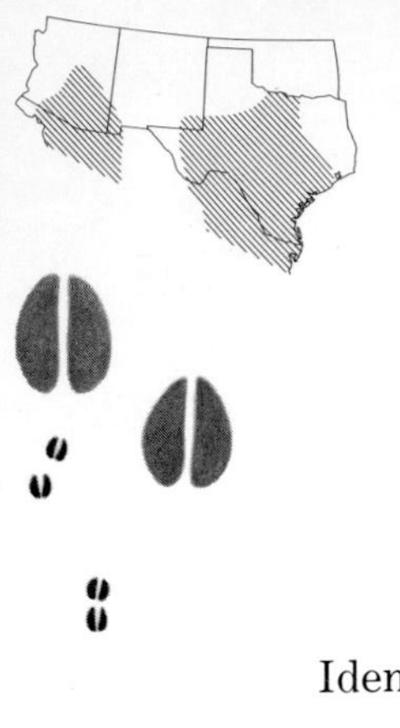

Native American relatives of pigs, peccaries travel in bands of up to 30, feeding on pricklypear cactus, the fruits of desert plants, and any mice, lizards, or insects they happen to find. Sometimes they move up into oak woodlands on slopes above the desert, where they forage on acorns. They are quiet during the midday heat and active in early morning and late afternoon. Peccaries are shy, but they have sharp tusks and can be dangerous if cornered. Most young peccaries are born in summer and are reddish or yellowish at birth. They can follow their mother the day after they are born.

Identification	34–37″. A small, coarse-haired, piglike animal with a long, narrow muzzle and a short tail. Grizzled grayish or blackish, tinged with yellow on sides of face, and with an indistinct pale "collar" across shoulders.
Similar Species	Unmistakable.
Habitat	Arid rocky or brushy country and deserts.
Range	S. Arizona and SW. New Mexico; also in central and S. Texas.

Pronghorn *Antilocapra americana*

With a top speed of 70 miles an hour, Pronghorns are the fastest animals in North America. They can see for long distances across the plains, and at the first sign of danger the long white hair on the rump flashes a signal to other members of a herd. Unlike Bison, sheep, and goats, Pronghorns drop the sheaths of their horns after mating in the fall. In the spring, one or two fawns are born and hidden in the grass for about a week, when they are strong enough to join the rest of the herd.

Identification 4–4½′ (male larger than female). A long-legged, deerlike animal with black, backward-curving horns, a foot or more long and with short forward prong in bucks, shorter and usually without prong in does. Tan or reddish with white underparts, flanks, and rump patch.

Similar Species Deer larger, without white patches, and with many-pronged antlers.

Habitat Grassy plains and sagebrush country.

Range SE. Oregon, Idaho, and Prairie Provinces south to Arizona and Texas.

White-tailed Deer *Odocoileus virginianus*

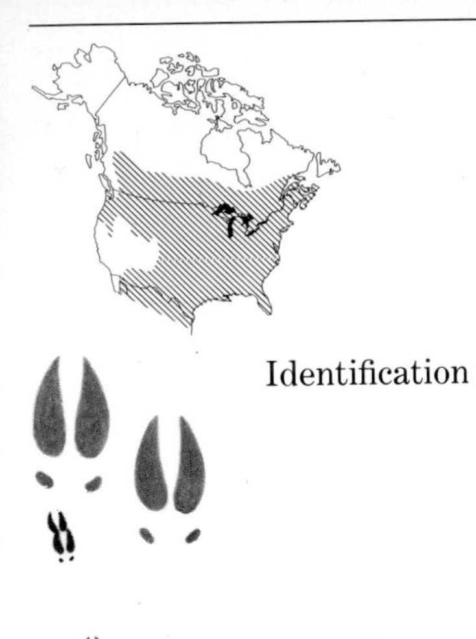

With hunting carefully regulated and plenty of suitable habitat, this graceful deer is now abundant in most parts of its range. Indeed, in much of eastern North America it lives too close to houses to be hunted. The spotted fawns, born in spring and usually in pairs, are weaned when they are four months old.

Identification — 4½–7′ (male larger than female). A slender deer with antlers having single main beam curved forward, a small tine over brow, and unbranched tines behind. Tan or reddish-brown in summer, grayer in winter, with white belly, throat, band over nose, and eye ring. Tail brown above with white border and white below; white underside of tail often lifted in alarm.

Similar Species — Mule Deer has larger ears, antlers with 2 equal branches, and black-tipped tail. Elk larger and darker.

Habitat — Woodlands, thickets, and suburban areas.

Range — S. Canada and continental United States except for most of California, Utah, N. Arizona, SW. Colorado, and NW. New Mexico.

Mule Deer *Odocoileus hemionus*

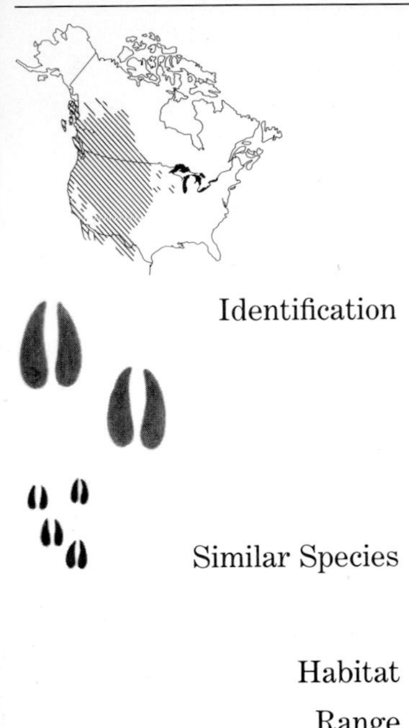

Named for their large, mobile ears, Mule Deer move about in small groups consisting of a doe, her fawns, and her year-old young. Bucks travel alone, but may form herds after the fall rutting season. In winter both sexes may gather in a "yard" where food can be found by scraping away the snow.

Identification 4–6½′ (male larger than female). A slender deer with large ears and antlers having 2 equal branches, each forking into 2 tines. Reddish-brown or yellowish-brown in summer, grayer in winter, with white throat and rump patch, whitish or tan underparts. Tail tipped black and usually white above (blackish or brown above in a form called the Black-tailed Deer of Pacific Northwest).

Similar Species White-tailed Deer has smaller ears, antlers with single main beam curving forward, no black tail tip. Elk larger and darker.

Habitat Forests, mountains, and grassy areas.

Range W. Canada and W. United States east to Manitoba, Wisconsin, and W. Texas and south into Mexico.

Elk *Cervus elaphus*

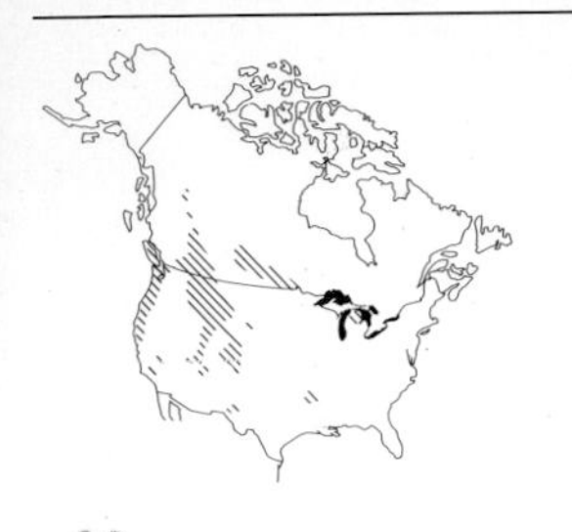

In the West the Elk, or Wapiti, spends the summer in alpine meadows and descends to the lowlands for the winter. During the fall rutting season, a male's loud, whistling challenge carries for a mile or more through the frosty air. A single spotted fawn is born in May or June after the adults have returned to the highlands. The name Wapiti, from an Algonquian word meaning "white," refers to the pale rump of this majestic deer.

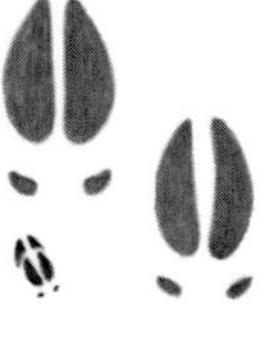

Identification 7–10′ (male larger than female). A large, robust deer with large, spreading antlers in males; light brown or dark brown above, darker below with pale rump patch and tail; dark reddish-brown mane on neck in males.

Similar Species White-tailed and Mule deer much smaller. Moose larger, dark brown with large muzzle and massive, flattened antlers. Caribou lacks white rump patch, has brow tines.

Habitat Forests, mountain meadows, and grasslands.

Range S. British Columbia and Manitoba south to N. California, Arizona, and New Mexico; scattered herds elsewhere. Formerly east to Massachusetts.

Caribou *Rangifer tarandus*

The migrations of tundra-dwelling Caribou provide one of the world's greatest wildlife spectacles. Constantly on the move, herds of up to 100,000 begin a trek in late winter from the stunted spruce and fir forest to the open tundra, where calves are born and the herd fattens in preparation for the fall migration back to the shelter of the trees. Farther south, other Caribou never leave the forest and live in smaller herds that make shorter migrations.

Identification: 4½–8′ (male larger than female). A large, shaggy-haired deer with long, curved antlers with many-tined branches and flat tine over brow. Color varies from brown with whitish throat, neck, and underparts (most of range) to nearly all-white (Arctic islands).

Similar Species: Elk has pale rump patch, lacks brow tines and white throat and neck.

Habitat: Tundra and northern coniferous forests.

Range: Alaska and Canada south to Idaho, north shore of Lake Superior, Quebec, and Newfoundland.

Moose *Alces alces*

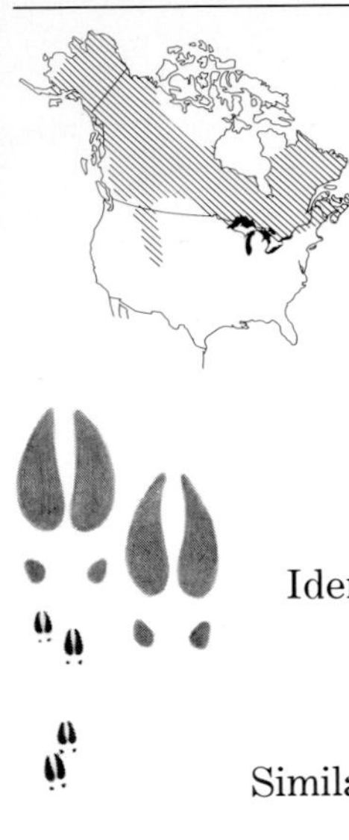

The largest deer in the world, the Moose is a forest dweller that browses on the twigs, bark, and buds of many shrubs and trees in winter and mainly on willows and aquatic plants in summer. In the warmer months, Moose are almost aquatic themselves. They are excellent swimmers and waders and often submerge themselves completely to avoid black flies. Calves are born in late spring and remain with their mother for a year. Moose can be very shy and secretive, but a bull may charge if disturbed during the fall rutting season.

Identification	7–10′ (male larger than female). A very large, long-legged, horselike deer with huge, flattened antlers, heavy muzzle, and dewlap under throat. Dark brown with paler legs.
Similar Species	Elk smaller with pale rump patch, more slender antlers.
Habitat	Forests, thickets, and swamps.
Range	Alaska and Canada south to Utah, NW. Colorado, the Great Lakes region, N. New York, and N. New England.

Bison *Bison bison*

About 60,000,000 "Buffalo" once inhabited grasslands and forests from Oregon and New Mexico east to New York, Maryland, and Florida, and north perhaps to Alaska. But only on the Great Plains was there enough grass for the huge herds early explorers found. Cows produce a calf every year, and Native Americans had little impact on the Bison's numbers. Rifles and shotguns changed the balance, and by 1900 the population had dropped to only about 1,000. Today with protection, there are about 30,000 Bison in North America.

Identification 10–12½′ (male); 7–8′ (female). A large, heavy-bodied, shaggy, dark brown animal with massive head; short, curved horns; beard under chin; and large hump over shoulders.

Similar Species Unmistakable.

Habitat Prairies, plains, and forests.

Range Wild herds now occur only at Yellowstone National Park in Wyoming and at Wood Buffalo National Park in Alberta and Northwest Territories.

Bighorn Sheep *Ovis canadensis*

The first snows of fall drive the Bighorn Sheep down from the highest meadows to slopes where the wind keeps the ground clear for foraging. It is here that the rams stage their annual sparring matches, butting their foreheads together with a crack that can be heard for a mile. The winners mate with the ewes, and the following spring, when the herds are back in the high country, a single lamb is born on a narrow, protected ledge.

Identification 4–6′ (male larger than female). A stocky wild sheep with short hair. Male (ram) has heavy, coiled horns; female (ewe) more slender, curved horns. Dark brown fading to tan in late winter or tan year-round in deserts. Whitish rump, muzzle, and eye patch.

Similar Species Dall's Sheep (*O. dalli*) smaller and white, gray, or blackish with more slender horns; found in Alaska and NW. Canada. Mountain Goat white, with slender horns.

Habitat Rocky cliffs and mountain meadows.

Range British Columbia, Alberta, and Montana south to SE. California, Arizona, and New Mexico.

Mountain Goat *Oreamnos americanus*

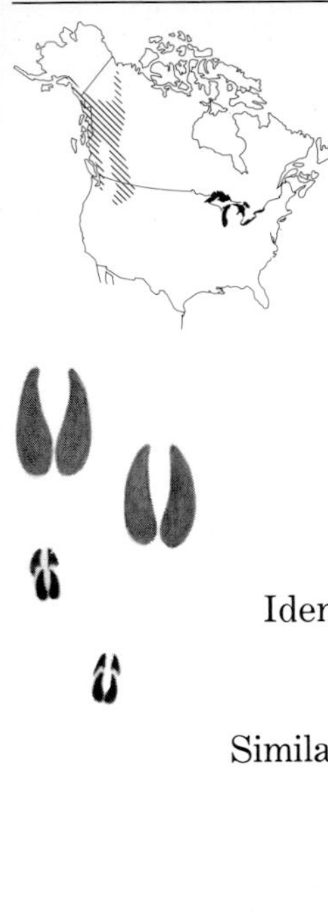

The highest and most forbidding crags are the home of the Mountain Goat. Here these agile grazers pick their way along narrow ledges and feed on tiny alpine plants. They seem almost like acrobats as they climb about in this treacherous habitat, but occasionally a Mountain Goat loses its balance, and avalanches are a greater danger than any predator. Their horns are more delicate than those of the Bighorn Sheep, and during the fall rutting season male goats spar only gently or merely threaten one another. After the snow has melted in late spring, one or two kids are born. They remain with their mother for a year.

Identification	4–6′ (male larger than female). A shaggy, white or yellowish-white wild goat with slender, slightly curved, black horns and a short beard under throat.
Similar Species	Bighorn Sheep darker with heavier, more curved horns.
Habitat	High mountain ledges and peaks.
Range	Alaska and W. Canada south to Washington, Idaho, and Montana. Introduced in Wyoming and Colorado.

Guide to Orders and Families

Mammals are classified in groups called orders and families. Knowing the general features of these groups is often helpful in identifying species.

Armadillos These animals belong to the small order Xenarthra, which also includes the anteaters and sloths of the tropics. All share features that are not visible on the outside of the body. The armadillos (family Dasypodidae) are unique in having the skin thickened into a bony armor. They have numerous peglike teeth for feeding on insects, stout claws for digging, and long ears.

Bats The order Chiroptera includes the only mammals capable of flying. Their front limbs serve as wings, with long, slender bones supporting a thin membrane that also connects the short hind legs and the tail. They have large ears that can pick up high-pitched sounds, and many sharp teeth. Most American bats belong to the family Vespertilionidae, but a few are free-tailed bats (family Molossidae), which differ in having the tail extending well beyond the membrane between the legs.

Carnivores Flesh-eating mammals of the order Carnivora all have sharp canine teeth to aid them in grasping their prey, and claws on their feet.

Foxes and Wolves Foxes and wolves (family Canidae) are wild relatives of dogs, with strong teeth, long, slender legs adapted for running, narrow muzzles, and large ears.

Bears and Raccoons The biggest carnivores are the bears (family Ursidae), heavy-set animals with large feet and long, powerful claws; they supplement their diet with berries and roots. The Raccoon and Ringtail (family Procyonidae) have pointed muzzles and banded tails; like the bears, they eat berries and other fruit as well as small animals.

Weasels and Cats Weasels and their allies (family Mustelidae) are a varied group that includes agile predators like the River Otter, Mink, Fisher, and Long-tailed Weasel, as well as stockier and slower-moving hunters such as the Badger and Wolverine, and the short-legged, black-and-white skunks that rely on an offensive spray for protection from their enemies. All have needle-sharp teeth and blunt claws for digging. The Bobcat, Lynx, and Mountain Lion are all cats (family Felidae), mainly nocturnal hunters that stalk their prey and subdue it with razor-sharp, retractile claws and sharp teeth.

Seals and Sea Lions Some carnivores have taken to the sea. The sea lions and fur seals (family Otariidae) have visible external ears and hind flippers that bend forward to assist in moving about on land. Most of them breed in large colonies on islands. The Harbor Seal is one of the hair seals (family Phocidae), streamlined swimmers that lack external ears and have hind flippers that are used like fins and cannot be bent forward.

Hoofed Mammals

There are four families of hoofed mammals or ungulates in North America. All have hoofs on two toes, rather than the single hoof that distinguishes horses, and they usually travel in herds.

Peccary

The Collared Peccary (family Tayassuidae), a relative of the pigs, is a stocky animal with a long muzzle, short, straight tusks, and a short tail.

Deer and Pronghorn

All male deer (family Cervidae) are easily recognized by their branching antlers, which are shed after every mating season; in some species the females also have antlers. The fleet-footed Pronghorn is the only member of the family Antilocapridae. Both sexes have a pair of curved horns made of bone, with an outer sheath that is shed each year; the white fur on the rump is raised as an alarm signal when danger is sighted.

Bison, Sheep, and Goats

The Bison, Bighorn Sheep, and Mountain Goat (family

Bovidae) are relatives of cattle; both sexes have permanent, unbranched horns of bone.

Marsupials The Virginia Opossum (family Didelphidae) is our only member of the order Marsupialia, which also includes the kangaroos of Australia. These are animals whose young are premature at birth, and complete their development in a fur-lined pouch on the mother's belly. Opossums have a long snout with many unspecialized teeth, coarse fur, and a long, naked tail.

Rabbits and Pikas The order Lagomorpha includes the rabbits and hares (family Leporidae) and the pikas (family Ochotonidae). They resemble rodents, but instead of a single pair of gnawing teeth or incisors on the upper jaw, they have two pairs, the second very small. The tail is always short, and they have five toes on both front and hind feet. Rabbits and hares have long ears, and the hind legs are longer than the front legs and adapted for running and bounding. Pikas are small rock-inhabiting animals with rounded ears and short legs.

Rodents Nearly half of all North American mammal species are members of the order Rodentia. Their distinguishing feature is a single pair of chisel-like teeth at the tip of the upper jaw. Although they all have these gnawing teeth, rodents are otherwise a varied group.

Mountain Beaver and Squirrels The secretive Mountain Beaver (family Aplodontidae) is a robust, burrowing animal with almost no tail. Squirrels and their allies make up the family Sciuridae. Many have bushy tails, and except for the flying squirrels they are active during the day. They range from the familiar tree squirrels to burrowing prairie dogs and rock-dwelling marmots.

Pocket Gophers The pocket gophers (family Geomyidae) are rodents that have adopted the underground lifestyle of moles, with very small ears and stout claws for digging. They feed on tubers and roots rather than on insects and earthworms as moles do.

Kangaroo Rats Nocturnal inhabitants of deserts and dry plains, kangaroo rats (family Heteromyidae) have small front feet but their hind legs are very long and adapted for leaping, and the long tail serves as a balance.

Beaver The Beaver, our largest rodent and the only North American member of the family Castoridae, is well known for its flattened tail, its appetite for the soft bark of twigs, and its habit of damming streams to make ponds.

Voles and Mice Voles, harvest mice, woodrats, White-footed and Deer mice, and muskrats (family Cricetidae), and the introduced rats and House Mouse (family Muridae) form two groups that differ in the shape of their grinding teeth. Outwardly they are very similar, with slender, often hairless tails and rather large ears. The jumping mice (family Zapodidae) are a small group of agile mice with long legs and long tails that jump swiftly through damp meadows or vegetation on the forest floor.

Porcupine and Nutria The tree-dwelling Porcupine, famous for its quills, is the only North American member of the family Erithizontidae. Although the introduced Nutria looks very much like a large Muskrat, it differs in the structure of the skull and is placed in the family Myocastoridae.

Shrews and Moles The order Insectivora includes the shrews (family Soricidae) and moles (family Talpidae), small animals with silky or velvety fur, very small eyes and ears, a pointed snout that extends well beyond the mouth, and many sharp teeth. Both are secretive, shrews keeping out of sight in dense vegetation or under leaf litter, and moles burrowing in the ground. Moles have poor vision but very keen hearing.

Orders of Mammals

Armadillos
page 90

Bats
pages 104–108

Carnivores
pages 110–156

Hoofed Mammals
pages 158–176

Marsupials
page 88

Rabbits and Pikas
pages 18–24, 56

Rodents
pages 26–54, 58–86, 96

Shrews and Moles
pages 92–94, 98–102

Index

Numbers in italics refer to mammals mentioned as similar species.

A
Alces alces, 170
Alopex lagopus, 110
Antilocapra americana, 160
Aplodontia rufa, 82
Armadillo, Nine-banded, 90

B
Badger, 148
Bassariscus astutus, 144
Bat, Brazilian Free-tailed, 108
Bat, Hoary, *106*
Bat, Indiana, *104*
Bat, Red, 106
Bear, Black, 152
Bear, Grizzly, 154
Bear, Polar, 156
Beaver, 84
Beaver, Mountain, 82
Bison, 172
Bison bison, 172
Blarina brevicauda, *102*
Bobcat, 124

C
Canis latrans, 118
Canis lupus, 120
Canis rufus, *120*
Caribou, 168
Castor canadensis, 84
Cervus elaphus, 166
Chipmunk, Cliff, *54*
Chipmunk, Eastern, 52
Chipmunk, Least, 54
Clethrionomys gapperi, 60
Clethrionomys occidentalis, *60*
Condylura cristata, 92
Cottontail, Desert, *24*
Cottontail, Eastern, 20
Coyote, 118
Cryptotis parva, 102
Cynomys leucurus, *28*
Cynomys ludovicianus, 28

D
Dasypus novemcinctus, 90
Deer, Mule, 164
Deer, White-tailed, 162
Didelphis virginiana, 88
Dipodomys ordii, 70

E
Elk, 166
Enhydra lutris, *132*
Erethizon dorsatum, 86
Ermine, *134*
Eumetopias jubatus, *130*

F
Felis concolor, 126

Felis lynx, 122
Felis rufus, 124
Fisher, 136
Flying Squirrel, Northern, *44*
Flying Squirrel, Southern, 44
Fox, Arctic, 110
Fox, Channel Islands Gray, *114*
Fox, Gray, 114
Fox, Kit, 116
Fox, Red, 112
Fox, Swift, *116*

G
Geomys bursarius, 96
Glaucomys sabrinus, *44*
Glaucomys volans, 44
Goat, Mountain, 176
Gopher, Plains Pocket, 96
Ground Squirrel, California, 46
Ground Squirrel, Cascade Golden-mantled, *50*
Ground Squirrel, Golden-mantled, 50
Ground Squirrel, Thirteen-lined, 48
Gulo gulo, 150

H
Halichoerus grypus, *128*
Hare, Snowshoe, 18
Harvest Mouse, Eastern, *72*
Harvest Mouse, Western, 72

J
Jackrabbit, Black-tailed, 22
Jackrabbit, White-tailed, *22*
Jumping Mouse, Meadow, 62
Jumping Mouse, Western, *62*

K
Kangaroo Rat, Ord's, 70

L
Lasiurus borealis, 106
Lasiurus cinereus, *106*
Lepus americanus, 18
Lepus californicus, 22
Lepus townsendii, *22*
Lion, Mountain, 126
Lutra canadensis, 132
Lynx, 122

M
Marmot, Hoary, 30
Marmot, Olympic, *30*
Marmot, Yellow-bellied, 32
Marmota caligata, 30
Marmota flaviventris, 32
Marmota monax, 26
Marmota olympus, *30*
Marten, *136*
Martes americana, *136*
Martes pennanti, 136
Mephitis mephitis, 140
Microtus pennsylvanicus, 58
Mink, 138
Mole, Broad-footed, *94*
Mole, Eastern, 94
Mole, Star-nosed, 92
Moose, 170
Mountain Beaver, 82
Mountain Lion, 126
Mouse, Deer, 66
Mouse, Eastern Harvest, *72*
Mouse, House, 64
Mouse, Meadow Jumping, 62
Mouse, Western Harvest, 72
Mouse, Western Jumping, *62*
Mouse, White-footed, 68
Muskrat, 80
Muskrat, Round-tailed, *80*
Mus musculus, 64
Mustela erminea, *134*
Mustela frenata, 134
Mustela vison, 138
Myocastor coypus, 78
Myotis, Little Brown, 104
Myotis lucifugus, 104
Myotis sodalis, *104*

N
Neofiber alleni, *80*
Neotoma floridana, 74
Nutria, 78

O
Ochotona collaris, *56*
Ochotona princeps, 56
Odocoileus hemionus, 164
Odocoileus virginianus, 162
Ondatra zibethicus, 80
Opossum, Virginia, 88
Oreamnos americanus, 176
Otter, River, 132
Otter, Sea, *132*
Ovis canadensis, 174
Ovis dalli, *174*

P
Peccary, Collared, 158
Peromyscus leucopus, 68
Peromyscus maniculatus, 66
Phoca vitulina, 128
Pika, 56
Pika, Collared, *56*
Pocket Gopher, Plains, 96
Porcupine, 86
Prairie Dog, Black-tailed, 28
Prairie Dog, White-tailed, *28*
Procyon lotor, 146
Pronghorn, 160

R
Rabbit, Brush, 24
Rabbit, Marsh, *20*
Raccoon, 146
Rangifer tarandus, 168
Rat, Black, *76*
Rat, Norway, 76
Rat, Ord's Kangaroo, 70
Rattus norvegicus, 76
Rattus rattus, *76*
Reithrodontomys humulis, *72*
Reithrodontomys megalotis, 72
Ringtail, 144

S
Scalopus aquaticus, 94
Scapaneus latimanus, *94*
Sciurus aberti, 42
Sciurus arizonensis, *42*
Sciurus carolinensis, 38
Sciurus griseus, 40
Sciurus niger, 34
Sea Lion, California, 130
Sea Lion, Northern, *130*
Seal, Gray, *128*
Seal, Harbor, 128
Sheep, Bighorn, 174

Sheep, Dall's, *174*
Shrew, Least, 102
Shrew, Masked, 98
Shrew, Short-tailed, *102*
Shrew, Vagrant, 100
Skunk, Eastern Spotted, *142*
Skunk, Striped, 140
Skunk, Western Spotted, 142
Sorex cinereus, 98
Sorex vagrans, 100
Spermophilus beecheyi, 46
Spermophilus lateralis, 50
Spermophilus saturatus, *50*
Spermophilus tridecemlineatus, 48
Spilogale gracilis, 142
Spilogale putorius, *142*
Squirrel, Abert's, 42
Squirrel, Arizona Gray, *42*
Squirrel, California Ground, 46
Squirrel, Cascade Golden-mantled Ground, *50*
Squirrel, Douglas', *36*
Squirrel, Fox, 34
Squirrel, Golden-mantled Ground, 50
Squirrel, Gray, 38
Squirrel, Northern Flying, *44*
Squirrel, Red, 36
Squirrel, Southern Flying, 44
Squirrel, Thirteen-lined Ground, 48
Squirrel, Western Gray, 40
Sylvilagus audubonii, *24*
Sylvilagus bachmani, 24
Sylvilagus floridanus, 20
Sylvilagus palustris, *20*

T

Tadarida brasiliensis, 108
Tamias dorsalis, *54*
Tamias minimus, 54
Tamias striatus, 52
Tamiasciurus douglasii, *36*
Tamiasciurus hudsonicus, 36
Taxidea taxus, 148
Tayassu tajacu, 158

U

Urocyon cinereoargenteus, 114
Urocyon littoralis, *114*
Ursus americanus, 152
Ursus arctos, 154
Ursus maritimus, 156

V

Vole, Meadow, 58
Vole, Southern Red-backed, 60
Vole, Western Red-backed, *60*
Vulpes macrotis, 116
Vulpes velox, *116*
Vulpes vulpes, 112

W

Weasel, Long-tailed, 134
Wolf, Gray, 120
Wolf, Red, *120*
Wolverine, 150
Woodchuck, 26
Woodrat, Eastern, 74

Z

Zalophus californianus, 130
Zapus hudsonius, 62
Zapus princeps, *62*

Photographers

Amwest
Y. Momatiuk (87), Mark Newman (155), Leonard Lee Rue III (115), Charles G. Summers, Jr. (145)

Roger W. Barbour (97, 101)
Fred Bruemmer (111)
Larry R. Ditto (23, 79, 135, 161, 163)
Harry Engels (85, 119, 177)
Jeff Foott (47, 165, 173, 175)
Francois Gohier (37)
Chuck Gordon (139)
William Grenfell (31)
G. C. Kelley (19, 55, 99, 125, 127, 159, 167)
Zig Leszczynski (93, 95, 147)
John R. MacGregor (65, 77, 103, 107)
Joe and Carol McDonald (45, 89, 113)
C. Allan Morgan (129, 131, 133)

National Audubon Society Collection/Photo Researchers, Inc.
Tom and Pat Leeson (83), Anthony Mercieca (73)
James F. Parnell (39, 43, 49, 53, 61, 63, 67)
Carroll W. Perkins (51)

Photo/Nats
Gay Bumgarner (35)

Rod Planck (29)
Helen Rhode (169)

Root Resources
Kenneth W. Fink (117), Anthony Mercieca (143)

Tom Stack & Associates
Rod Allin (149), Mary Clay (81), Kerry T. Givens (105), Thomas Kitchin (41, 121, 141, 153), Robert McKenzie (25), Gary Milburn (123, 151), Brian Parker (157), Milton Rand (137), John Shaw (57)

Alvin E. Staffan (27, 59, 69, 75)
Robert Villani (171)
Larry West (91)
Jack Wilburn (21, 33)
Dale and Marian Zimmerman (71, 109)

Cover Photograph

Elk by G.C. Kelley

Illustrators

Range maps and silhouettes by Paul Singer
Tracks by Dot Barlowe

Cover photograph:
Elk by G. C. Kelley

Title page: Gray Squirrel by James F. Parnell

Spread (16–17): Red Fox by Jeff Lepore/Photo Researchers, Inc.

Staff for this book

Publisher: Andrew Stewart
Editor-in-Chief: Gudrun Buettner
Executive Editor: Susan Costello
Managing Editor: Jane Opper
Assistant Editor: Amy K. Hughes
Production Manager: Helga Lose
Production: Gina Stead-Thomas, Chris Adams
Art Director: Carol Nehring
Art Associate: Ayn Svoboda
Art Assistant: Cheryl Miller
Picture Library: Edward Douglas

Original series design by
Massimo Vignelli

All editorial inquiries on this title should be addressed to:
Pocket Guides
P. O. Box 479
Ascutney, VT 05030
editors@thefieldguideproject.com

To purchase this book or other National Audubon Society illustrated nature books, please contact:
Alfred A. Knopf, Inc.
1745 Broadway
New York, NY 10019
(800) 733-3000
www.aaknopf.com

NATIONAL AUDUBON SOCIETY

The mission of NATIONAL AUDUBON SOCIETY *is to conserve and restore natural ecosystems, focusing on birds, other wildlife, and their habitats for the benefit of humanity and the earth's biological diversity.*

One of the largest environmental organizations, AUDUBON has 550,000 members, 100 sanctuaries and nature centers, and 508 chapters in the Americas, plus a professional staff of scientists, educators, and policy analysts.

The award-winning *Audubon* magazine, sent to all members, carries outstanding articles and color photography on wildlife, nature, the environment, and conservation. Audubon also publishes *Audubon Adventures*, a children's newspaper reaching 450,000 students. Audubon offers nature education for teachers, families, and children through ecology camps and workshops in Maine, Connecticut, and Wyoming, plus unique, traveling undergraduate and graduate degree programs through *Audubon Expedition Institute.*

AUDUBON sponsors books, on-line nature activities, and travel programs to exotic places like Antarctica, Africa, Baja California, and the Galápagos Islands. For information about how to become an Audubon member, to subscribe to *Audubon Adventures*, or to learn more about our camps and workshops, please contact:

AUDUBON
225 Varick Street, 7th Floor
New York, NY 10014
(212) 979-3000 or (800) 274-4201
www.audubon.org